Terrorism, Security, and Computation

Series Editor

V. S. Subrahmanian, College Park, USA

For further volumes:
http://www.springer.com/series/11955

Patricia Taft • Nate Haken

Violence in Nigeria

Patterns and Trends

Springer

Patricia Taft
The Fund for Peace
Washington, DC, USA

Nate Haken
The Fund for Peace
Washington, DC, USA

The Authors would like to thank the Foundation for Partnership Initiatives in the Niger Delta, Nigeria Watch, FFP's UNLocK, CFR's Nigeria Security Tracker, WANEP Nigeria, CSS/ETH Zurich, ACLED, and The Gadfly Project.

ISSN 2197-8778   ISSN 2197-8786   (electronic)
Terrorism, Security, and Computation
ISBN 978-3-319-36019-5   ISBN 978-3-319-14935-6   (eBook)
DOI 10.1007/978-3-319-14935-6

Acknowledgments

The Authors would like to thank the organizations and initiatives that work tirelessly to collect data on issues of peace and conflict in Nigeria used in this book (Nigeria Watch, FFP's UNLocK, CFR's Nigeria Security Tracker, WANEP Nigeria, CSS/ETH Zurich, and ACLED). We also want to thank The Gadfly Project for working with us to design the Peace Map so that the data from those sources could be triangulated, cross-validated, and visualized. Most of all we want to thank the Foundation for Partnership Initiatives in the Niger Delta for making the Peace Map possible, and for the work you do in applying this analysis for peacebuilding and development on the ground. Many thanks to the research support provided for this project by Sofia Scott, Paul Friesen, Jenna Greene, Tuba Dokur, Ellen Galdava, Katherine Carter, Valentin Robiliard, Laura Brisard, Charles Fiertz, Alexandra Kelly, Jennifer Lowry, Marcela Aguirre, Kadeem Khan, Alexander Young, Ania Skinner, Katie Cornelius, and Hannah Blyth.

Contents

Acronyms

ACLED	Armed Conflict Location & Event Data Project
ACN	Action Congress of Nigeria
ACO-MORAN	Amalgamated Commercial Motorcycle Owner and Riders Association of Nigeria
AD	Alliance for Democracy
AIYO	Association of Igbo Youths
ANPP	All Nigeria Peoples Party
ANSARU (or JAMBS)	Jamā'atu Anṣāril Muslimīna fī Bilādis Sūdān
AP	African Petroleum
APC	All Progressives Congress
APGA	All Progressive Grand Alliance
APP	All People's Party
ASTA	Anambra State Traffic Agency
ASUU	Academic Staff Union of Nigerian Universities
BVYG	Borno Vigilance Youths Group
CFR	The Council on Foreign Relations
CPC	Conscience People's Congress
CSS/ETH Zurich	Center for Strategic Studies/Eidgenössische Technische Hochschule Zürich (Swiss Federal Institute of Technology Zurich)
CYGG	Coalition of Youth for Good Governance
EFCC	Economic and Financial Crimes Commission
FFP	The Fund for Peace
ICJ	International Court of Justice
ICPC	Independent Corrupt Practices Commission
IED	Improvised Explosive Device
INEC	Independent National Electoral Commission
JAMDS	Joint Admission and Matriculations Board
JAMBS (or ANSARU)	Jamā'atu Anṣāril Muslimīna fī Bilādis Sūdān
JAS	Jamā'at Ahl as-Sunnah lid-da'wa wal-Jihād (Boko Haram)

JTF	Joint Task Force
LASTMA	Lagos State Traffic Management Authority
LGA	Local Government Area
LP	Labour Party
MAPOLY	Moshood Abiola Polytechnic
MASSOB	Movement for the Actualization of the Sovereign State of Biafra
MEND	Movement for the Emancipation of the Niger Delta
NAFDAC	National Agency for Food and Drug Administration Control
NANS	National Association of Nigerian Students
NDSF	Niger Delta Subterranean Force
NJC	National Judicial Council
NLC	Nigeria Labour Congress
NNPC	Nigerian National Petroleum Corporation
NSCDC	Nigeria Security and Civil Defence Corps
NULGE	Nigeria Union of Local Government Employees
NUPENG	Nigeria Union of Petroleum and Natural Gas
NURTW	National Union of Road Transport Workers
NYSC	National Youth Service Corps
OPC	Oodua People's Congress
PDP	People's Democratic Party
PHCN	Power Holding Company of Nigeria
PIND	Foundation of Partnership Initiatives in the Niger Delta
PPA	Progressive People's Alliance
PTI	Petroleum Training Institute
RTEAN	Road Transport Employers Association of Nigeria
SARS	Special Anti-Robbery Squad
TOAN	Truck Owners Association of Nigeria
UN	United Nations
UNESCO	United Nations Educational, Scientific, and Cultural Organization
UNIPORT	University of Port Harcourt
UNIUYO	University of Uyo
VGN	Vigilante Group of Nigeria
ZSIEC	Zamfara State Independent Election Commission

Chapter 1
Introduction

Nigeria's Fourth Republic, which began in 1999 with the election of Olusegun Obasanjo followed by Umaru Yar'Adua and Goodluck Jonathan, has been fraught with security challenges, including spikes in communal and sectarian violence in the Middle Belt, militancy and criminality in the Niger Delta, and insurgency in the Northeast. Those with a stake in peace and security in Nigeria, whether they be donor agencies, security services, civil society, government, private sector, or community leaders, all rely on data and indices for their baselines; site-selection; resource allocation; and monitoring and evaluation of their programs, projects, and activities. The level of analysis varies by stakeholder, some of whom are focused at the national level, while others are focused at the state, local government, or community levels. However, regardless of mandate, projects can easily be derailed if a systems lens is not applied in recognition of the fact that local level pressures can impact state and national trends, just as national trends can impact local patterns. An analytical focus on Boko Haram in the Northeast, for instance, to the exclusion of inter-communal conflict in the Middle Belt, can easily lead to misinterpretations and unpleasant surprises. A particular local government in the Niger Delta may have a long, peaceful history, but if the wider trends are not understood, traditional leaders may miss the early signs of conflict escalation until it is too late.

To say that conflict dynamics are complex is not especially new or insightful. However, the case of Nigeria may have more complexity than most. With over 170 million people, it is by far the most populous country in Africa and one of the world's most ethnic and culturally diverse. Lagos is among the largest cities in the world. Nigeria is among the top exporters of oil worldwide. Expressions of violence sometimes occur across a North/South dyad. Sometimes conflict appears religious or ethno-sectarian. Other times conflict is more overtly resource based. But faultlines are by no means immutable. Every time a bomb goes off or a new election takes place, the deck is reshuffled. Few incidents fit perfectly into any one model. And the dynamics vary from one state to the next, and from one Local Government Area (LGA) to the next.

© Springer International Publishing Switzerland 2015
P. Taft, N. Haken, *Violence in Nigeria*, Terrorism, Security,
and Computation, DOI 10.1007/978-3-319-14935-6_1

If peacebuilding is complex, and requires multi-stakeholder collaboration at the national, state, and local levels, a first step to an effective strategy is a catalog of patterns and trends. That is what this book tries to do. The Fund for Peace (FFP) and the Foundation for Partnership Initiatives in the Niger Delta (PIND) developed a web application (http://www.p4p-nigerdelta.org/peace-building-map) with data technologies provided by The Gadfly Project, to integrate data on peace and conflict in Nigeria from a wide range of sources, including ACLED, the Council on Foreign Relations' Nigeria Security Tracker, AOAV, FFP's UnLocK project, WANEP Nigeria, CSS/ETH Zurich's Energy Infrastructure Attack Database, NSRP Sources, and Nigeria Watch. Each data source has strengths and weaknesses. Integrating them all on a single platform allows for cross-validation and triangulation. Data is coded according to date, indicator, sub-indicator, region, state, and LGA. Visualizations include static heat maps, dynamic heat maps, clustered maps, tables, bar charts, and trend-lines. All data can be downloaded by the user for off-line analysis. Indicators were developed iteratively over the course of five years of participatory workshops in Liberia, Uganda, and Nigeria. Indicators continue to be adjusted as dynamics change and stakeholders identify new issues to be monitored. As of this writing, they include the following:

Conflict Indicators

- Human Rights

    - Child Abuse
    - Sexual Violence
    - Gender-Based Human Rights Violations
    - Media Freedom
    - Unlawful Arrest
    - Human Trafficking
    - Domestic Violence
    - Other Human Rights Violations

- Demographic Pressures

    - Land Competition/Cattle Rustling
    - Natural Disasters/Drought
    - Disease Outbreaks
    - Environmental Degradation
    - Alcoholism/Narcotic Abuse
    - Food Crisis

- Insecurity

    - Shootings/Killings
    - Abductions
    - Terrorism
    - Vigilante/Mob Justice
    - Violent Protest/Crackdown

- Armed Clashes
- Abuses by Public Security Forces
- Arms Proliferation
- Bank Robberies
- Ambushes
- Cross-Border Conflict
- Crime
- Domestic Violence Fatality
- Ritual Killings
- Attacks on Energy Infrastructure

- Economic Pressures

  - Inflation
  - Unemployment
  - Poverty
  - Labor Strikes
  - Illicit Economy/Corruption
  - Extortion/Racketeering
  - Insecurity Hurts Business

- Group Grievance/Collective Violence

  - Hate Speech
  - Ethnic/Religious Tension
  - Tension or Violence between Political Groups
  - Intra-Communal Tension or Violence
  - Gang Violence
  - Insurgency/Counter-Insurgency
  - Inter-Communal Tension or Violence

- Governance/Legitimacy

  - Public Security Forces Corruption
  - Government Corruption
  - Riots/Protests
  - Election Irregularities
  - Intimidation of Political Opponents
  - Unresolved/Delayed/Disputed Elections/Impeached Officials

- Public Services

  - Health System
  - Education System
  - Power Supply
  - Prison System
  - Roads/Infrastructure
  - Water and Sanitation

- Refugees/IDPs

  - Displaced by Violence
  - Displaced by Disaster
  - Displaced by Land Seizure
  - General Displacement Issues

Peace Agents (Areas of Focus)

- Economic Development
- Human Rights
- Peace/Conflict Mitigation
- Gender
- Youth
- Governance
- Community Mobilization
- Emergency Response
- Environment
- Agriculture
- Education
- Health
- Human Security
- Community Development
- Communications
- Children

As of this writing there are over 13,000 incidents of conflict risk and 400 Agents of Peace mapped to the platform from January 2009 to July 2014. Hundreds more are being added every month, making it a valuable tool for multi-stakeholder information sharing and collaboration, especially among those who have reliable internet access and the time and resources to spend on research and data analysis. Effective response requires the involvement of those who may not have internet or time. To reach them, The Fund for Peace (FFP) and the Foundation for Partnership Initiatives in the Niger Delta (PIND) began using the tool to produce brief conflict bulletins which are published on the Partners for Peace website here: http://www.p4p-nigerdelta.org/conflict-bulletins. This book is an expansion of those bulletins to include all Nigerian states with a focus on 2009–2013.

It is critical to note that if the goal is to understand conflict in Nigeria, the findings in this book are only one piece of the puzzle. Patterns and trends tell you very little about causes and solutions. They may help to dispel oversimplifications or to suggest hypotheses about root causes and triggers. But going beyond that requires a much deeper, qualitative assessment, ideally taking a participatory approach involving those directly impacted by the salient conflict issues of concern. To that end, FFP and PIND have been facilitating a series of workshops across Nigeria where conflict bulletins are interrogated by local stakeholders, and hotspots identified and analyzed using FFP's Conflict Assessment System Tool (CAST) methodology, for a better understanding of root causes. Finally, local stakeholders then devise concrete plans for conflict mitigation based on the quantitative data tallied in the bulletins, and the qualitative analysis done in the workshop. Data and bulletins are similarly used by working groups and roundtables at various levels. We hope that this book, which expands significantly on the bulletins that have been produced, will serve a similar purpose.

Scope and Limitations of the Data: Each data source has strengths and weaknesses. For the sake of this analysis, we used Nigeria Watch data to quantify incidents and fatalities (excluding automobile accidents and other accidents) because unlike some of the other sources integrated onto the platform, Nigeria Watch (a) covers all states and LGAs, (b) covers the entire 5-year period of concern, and (c) covers the widest range of criminal, political, sectarian, inter-personal, and communal conflict issues. Once identifying the hotspots and trends using Nigeria Watch data, we used all data sources to fill in the incidents and issues listed by state and LGA, thus ensuring a comprehensive overview in the pages below. The coding of each incident by date and location is cross-validated to the fullest extent possible. However, given slight discrepancies from one source to the next, it is possible that some details may be inaccurate. For example, in some cases the LGA may have been coded incorrectly by the original source, or the date of a particular incident may be approximated. It is also true that some indicators are under-reported even taking into consideration the triangulation undertaken here, a limitation that has been corroborated by local stakeholders in workshops held in the Niger Delta and the Middle Belt with participants from across the country. These limitations notwithstanding, this catalogue gives a good overview of patterns and trends as evidenced by a careful integration of over 11,000 incidents from 2009 to 2013, which can serve as a baseline from which to do additional research by those with a stake in peace and security in Nigeria.

In an attempt to better understand the various conflict ecosystems across the country of Nigeria, the book divides the 36 states (plus the Federal Capital Territory) into seven regions: Niger Delta, North Central, Middle Belt, Northeast, Northwest, Southwest, and South Central. These regions are not official administrative or geopolitical zones, but have been defined based on proximity, size, and general commonality regarding conflict issues. As you read these pages, however, you'll see that at times, issues transcend these regions and at other times issues vary dramatically within the regions themselves. The regional breakdown is just an attempt to take a broad brush to the research question before drilling down to the state and local levels. Within each state, we highlight issues in the top five most insecure LGAs as measured by reported incidents per capita over the 5 year period. In cases where a state has multiple adjacent LGAs with the same name (e.g., Warri South, Warri North, Warri Southwest) we aggregated the incidents to mitigate against the possibility of coding error on the part of the original source.

At the national level, according to Nigeria Watch data (findings which are corroborated by ACLED, the other source that covers the entire period) violence was much more severe in 2013 than in 2009. Nigeria Watch reports escalating violence every year, with approximately twice as many incidents and fatalities in 2013 than in 2009 (Fig. 1.1).

Armed Groups: Most of that escalation was in the Northeast region, but armed groups proliferate across the country. Aside from the *public security forces* such as the police, the Joint Task Force (JTF) in the Niger Delta and the Special Task Force (STF) in the Middle Belt, there are criminal gangs, vigilante groups, insurgent groups, ethnonationalist militias, and separatists. *Insurgent groups* include radical jihadists in the North such as Jamā'at Ahl as-Sunnah lid-da'wa wal-Jihād (JAS), commonly referred to as Boko Haram, and Jama'atu Ansarul Musliminia fi Biladis Sudan (JAMBS), commonly referred to as Ansaru. In the South, the main insurgent group was called the Movement for the Emancipation of the Niger Delta (MEND), a faction of which is the Niger Delta Strike Force (NDSF). *"Cult Groups"* is a generic term for confraternities and criminal gangs whose structures and leadership are at times leveraged for political intimidation or militancy. Examples of cult groups mentioned in this book include the Aye and the Eiye in the southwestern part of the country and the Black Axe, the Red Axe, and the Blue Beret throughout the South. Also active in the Niger Delta was the Degban wing of the Klansmen cult group, their sometime rivals, the Dewell wing of the Vikings cult group, the Icelanders, and the Greenlanders. Small criminal gangs include the Ade Basket Boys in Ondo State, the Kauraye in Katsina, and AJ1 and the Senior Boys in Lagos State. *Vigilante Groups* formed to fight crime and insecurity, often on behalf of an ethnic constituency include the Bakassi Boys in Abia and Imo, the Oodua People's Congress (OPC) in the Southwest, the Borno Vigilance Youth Groups (BVYG) in the Northeast and, more locally, the Ejilewe Ukwuagba Neighbourhood Security Committee, in Ebonyi. Ombatse is the *Ethnic Militia* of the Eggon in Nasarawa. *Separatist Groups* include the Movement for the Actualization of the Sovereign State of Biafra (MASSOB) in the South Central region and around Imo State (Fig. 1.2).

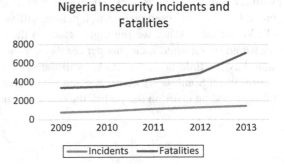

Fig. 1.1 Line Graph shows trend in # of incidents and fatalities in Nigeria

Timeline

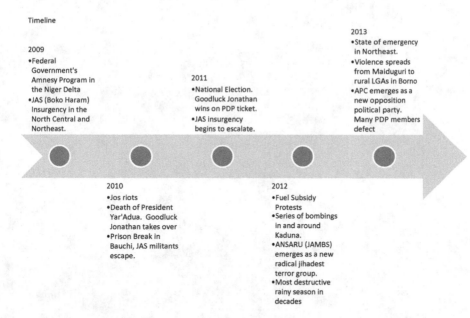

Fig. 1.2 Key events affecting security in Nigeria between the years of 2009 and 2013

The escalation in insecurity has not been evenly distributed across the regions, states, and local governments, however. Some areas have worsened, some improved, and some have experienced periodic spikes in violence followed by periods of calm. The following pages break the aggregate data down into its component parts for a better understanding of conflict and insecurity in Nigeria. We hope that readers find that it sheds light on a very complex issue and that they then take the next steps to identify and fill in gaps, qualitatively assess root causes and, ultimately, develop practical solutions for a more peaceful and secure Nigeria. We also encourage readers to do their own analysis using the interactive web map application at http://www.p4p-nigerdelta.org/peace-building-map.

Fig. 2.3 ...

Chapter 2
Niger Delta Overview

While the Niger Delta may be one of the most resource-rich regions in the world, it remains mired in cycles of conflict that perpetuate underdevelopment and threaten human security. As can be explored through a more in-depth analysis of the eight individual states of the Niger Delta that follows, the problems that plague the region are complex and often deeply entrenched. These issues include poor governance at all levels that is often exacerbated by a political culture that simultaneously centralizes power (and access to wealth) at the top yet relies more on informal networks than civic participation. In addition, the overall weak capacity of the state to protect its citizens at all levels has given rise to various non-state actors that have filled the vacuum, at a very violent and deadly cost. Compounding the general structural problems of the state and local level government, the cycles of violence that have so often plagued the Niger Delta are exacerbated greatly by the vastly uneven distribution of resources from the oil industry which are amassed at the top and rarely trickle down to benefit local communities. Adding to these pressures is the problem of extreme environmental degradation in a region where up to 60 % of people rely on natural resources for their overall livelihoods. Increased social fragmentation and long-simmering minority tensions also add layers of complexity to the security challenge in the Niger Delta, as various manifestations of these particular drivers of conflict often seem to appear out of nowhere and can quickly spiral into crises (Haken et al. 2013).

The oil-rich region of the Niger Delta is located in southern Nigeria at the delta of the Niger River. There are nine oil producing states, with a population of approximately 32 million, or 20 % of the overall population of the country, according to 2006 estimates. Demographically speaking, over 62 % of the population is under the age of thirty and there are more than forty different ethnic groups, with the Ijaw comprising the largest group. Access to higher education is among the lowest in all of Nigeria while infant mortality rates are among the highest. Despite the booming oil economy, unemployment remains rampant. When the high unemployment rates are factored in with a large youth population with little education and few opportunities, it translates into a restive population with few opportunities to pull themselves out of poverty (CIA

© Springer International Publishing Switzerland 2015
P. Taft, N. Haken, *Violence in Nigeria*, Terrorism, Security,
and Computation, DOI 10.1007/978-3-319-14935-6_2

World Factbook 2014). In addition, lack of legitimate employment opportunities and the marked uneven distribution of the region's resources have driven criminality and violence, with a history of youth recruitment into various armed groups. These armed gangs, often at the behest of more powerful political interests, have waged violent campaigns against the local government, the military, and international oil companies since shortly after independence in 1960. The overlapping of criminality with political ambitions is not unique to the Niger Delta, although it is certainly pronounced in the region and complicates peacebuilding and development efforts (Ibid).

The Role of Cults in the Niger Delta: While an exhaustive history of the Niger Delta is not possible here, in order to understand the various elements fueling instability, it is worthwhile to highlight some specific driving and complicating factors that manifest at the state level as well. First, as noted above, the high unemployment rate, particularly among youth, in the Niger Delta has long been a major source of instability as it has driven participation in criminal gangs, vigilante and militancy movements, and cult groups. Moreover, these groups have evolved along with the social and political dynamics of the Niger Delta and lines between them often overlap or blur. Further complicating analyses, the term "cult" is often very freely used in Nigeria, in particular in the media. The term "cult" may be used to describe any organized group where affiliation, ideology and operations are often kept secret. The term can also imply a religious or pseudo-religious dimension, often in the form of juju or occult practices (Haken et al. 2012).

Today, cults serve as an entryway into all kinds of criminality and violence, including militancy. Historically, however, these groups have played an important role in African societies, including in conflict and dispute resolution and the dispensation of justice. In Nigeria, cult groups were traditionally formed along ethnic lines comprising elders and adults. While cults exist throughout the country, they are most prevalent in the south, particularly the Niger Delta. Within this ever-changing social, political, and economic environment over the past 20 years, cults became most prevalent in tertiary education institutions, known as confraternities (Ibid). While universities in Nigeria have long been the home of highly active student associations, which traditionally provided memberships that lasted a lifetime and provided social and career benefits, they began to become more politicized and polarized along ideological lines in the early to mid-1970s, reflecting a wider societal shift along these lines. Around the mid-1980s, the nature of confraternities changed decidedly, from generally peaceful social and political clubs to restive and often violent gangs, known for terrorizing students as well as teachers and other public figures (Asuni 2009).

The original confraternities, founded at the University of Ibadan, were known as the Pyrates, which was later renamed the National Association of Sea Dogs, and the breakaway group, the Buccaneer Association. Later, in the mid-1970s, the University of Benin in Edo State became the epicenter of a confraternity movement known as the Neo-Black Movement of Africa, whose mission and ideological bent was centered on restoring and consolidating African culture and identity. In later years, notable confraternities which evolved into violent cult groups included the Klansmen

Konfraternity (KKK) in Cross Rivers State, as well as the Supreme Vikings Confraternity (SVC) and the Black Axe Confraternity, which have played active roles in the violence in Rivers and Bayelsa states. In particular, politicians in these two states have at times had strong linkages to these confraternities, especially in Rivers where some have been known to be affiliated directly or indirectly with the Vikings. In Rivers, at the University of Port Harcourt, clashes and rivalries between the Vikings and the KKK became common, and street gangs formed around the groups to support their violent undertakings, which included attacks on each other outside of the campus. Two notorious street gangs, the Deebam and the Icelanders, were formed to support the KKK and the Vikings, respectively. These groups widened their memberships to even younger boys, mostly teenagers, who were armed and partook in violent clashes with each other and in support of the larger confraternities. Beginning in the early 1990s, the Vikings and the KKK, and their affiliated street gangs, also became increasingly involved in the drug trade, with fierce battles waged over control of Port Harcourt market. In particular, wars between the KKK and the Icelanders over the drug market and territory were frequent and extremely bloody (Ibid).

Confraternity violence peaked in the 1990s, spreading to the streets and creeks through militarized wings which further multiplied through breakaway factions comprised largely of dropouts driven by economic incentives. In addition to the groups mentioned above, other urban cult gangs have their origins as groupings of young people that usually grew up together in impoverished neighborhoods and squatter camps. Over time, these groupings also evolved into street cult gangs dedicated to the protection of their members and territory, often through violence. Membership in cults also provided unemployed youth with economic opportunities in their communities, beyond the drug and weapons market. The most obvious choice for illicit economic activity was found in the illegal siphoning of petroleum, also known as bunkering (Francis 2011).

In addition to the drugs and oil bunkering businesses, other opportunities for illegal employment abound. During elections, for instance, powerful, local "godfathers" who were members or affiliated with specific groups, will often use young cult members for intimidation and political thuggery. Larger criminal gangs also use younger cult members for route protection and security for their illegal oil bunkering activities. It has been widely reported that oil companies, in order to secure their operations and provide safety for personnel, have been known to make direct payments to cult gangs, which, from an outside perspective, can be hard to identify and distinguish. Furthermore, beyond oil extortion and racketeering, cultists serve in a multitude of roles from bodyguards to hit men to organizers of prostitution and human trafficking rings. In recent years, cults have also been implicated in kidnapping and ransom syndicates (Haken et al. 2012). At all levels, including local and state government, a sophisticated system of bribery, patronage, and profit-sharing exists to support the activities of cults and other street gangs. In addition, it is believed that the involvement of foreign "clients," willing to buy bunkered oil at cheap rates, also contributes to the rapid growth and financial gains of individual groups.

Over the past decade, and sometimes with the direct support of the local and federal government, various state and non-state actors have been mobilized to try to counter the threat posed by cult groups. Particularly in Delta, Bayelsa and Rivers states, as evidenced by the number of violent incidents directly attributed to them, cult gangs are a serious problem. At times, attempts by the government to eradicate cults have, in several instances, succeeded in temporarily reducing violence only for it to flare up again, often much more dramatically. In addition, urban vigilante groups, formed to protect their neighborhoods from cult and criminal gangs, may have realized some success in combatting cultism locally but have also driven the problem to more rural areas. The prevalence and reach of cults in the Niger Delta points to several drivers of conflict that undermine human security, including the lack of opportunities for a gainful livelihood compounded by the inefficiency and corruption of state institutions, including the police and local government. The infiltration by cults and criminal gangs into local politics and the police underscores this problem. Additionally, cult gangs and the formation of vigilante groups also demonstrates the profound gaps in police capacity to deal with wholesale insecurity. The government, rather than rely on the police, routinely employs the military to deal with cult groups and criminal gangs, normally leading to not only an escalation of violence, but also high levels of civilian casualties and property damage (Asuni 2009).

In addition to confraternities and cult gangs, the Niger Delta also plays host to various other groups including advocacy and resistance organizations (including umbrella coalitions), militias, vigilante groups and, as mentioned above, various criminal gangs with some or no affiliation to cults. Noting again that an exhaustive history of such groups is not feasible within the parameters of this book, the origins and current roles of the main groups are fundamental to understanding the current context of the Niger Delta, as is a brief history of the grievances that helped drive their formation.

A Brief History of Armed Conflict in the Niger Delta: The history of armed conflict in the Niger Delta began in early 1966 and set the stage for many of the dynamics that would continue to play out in the region for coming decades. Motivated by the control for oil revenue, in February 1966, the Eastern Niger Delta region, consisting of Rivers and Bayelsa states, were declared independent by an Ijaw named Isaac Borow, the leader of the Niger Delta Volunteer Force. Initially called the Niger Delta People's Republic, later the Republic of Biafra, Boro called on international oil companies to negotiate directly with the Republic, rather than with the national government. Although this attempt at independence only lasted less than 2 weeks, it presaged the Biafran Civil War which would catapult the Niger Delta into a deadly civil war for almost 3 years, from May 1967 through January 1970. During this time period, the eastern region was broken into three states: Rivers, Bayelsa and Delta, with Rivers comprising the majority of the oil producing areas, along with the highest concentration of minority tribes. Also, both during and after the war, legislation including the 1969 Petroleum Decree and the 1978 Land Use Decree both gave control of all resources to the federal government while nationalizing all land under the administration of state and local authorities. While this move initially helped assuage tensions between the countries three main ethnic groups; the Hausa, Igbo and Yoruba, it simultaneously served to disenfranchise and marginalize many of the minorities of the Delta, who did not find their interests represented. Later, divided interests and lingering grievances would again fuel discord among the Hausa, Igbo and Yoruba as well.

In addition to a devastating civil war and the firm defeat of local aspirations for the control of oil revenues, environmental degradation also had devastating effects on livelihoods. In protest to both the degradation of local land as well as out of the perceived unfair distribution of oil revenue and marginalization, 1990 saw the creation of another significant group, the Movement for the Survival of the Ogoni People (MOSOP), led by Ken Saro-Wiwa. In the same year, MOPSOP introduced the Ogoni Bill of Rights, calling on the government to grant greater resource control to the people of Ogoni Land while acting to protect the local environment from destruction. While the movement itself was initially non-violent and utilized protests as its main form of political demonstration, the federal government often reacted with a heavy-handed response, often deploying the military to deal with demonstrators. Also during this time period, the region as a whole became more militarized, with an uptick in the flow of weapons fueling increased militancy. At the federal level, the 1993 dissolution of democratic institutions and imposition of military rule did little to discourage the growing trend of turning to violence to protest perceived injustices and resolve disputes throughout the Niger Delta. The execution of MOSOP leader, Ken Saro-Wiwa, along with eight others, further entrenched deep-seated grievances between the federal government, military and local populations. It also led to Nigeria's status as a pariah state, as international sanctions were imposed and nearly all development assistance to the Niger Delta halted (Francis 2011).

Throughout the 1980s and early 1990s, relationships continued to fracture between and within ethnic groups and communities as well as between oil companies and communities. In particular, the breakdown of inter-communal relationships and the erosion of traditional leadership also became more pronounced during this time period, and would only worsen in the future. The weakening of traditional leadership structures which, in part, began during the colonial period, was further exacerbated by the fierce competition for resources. The designation of "host communities," or those eligible for compensation from oil companies and potentially positioned for better employment opportunities, not only caused fragmentation and fierce competition between communities, but also within them. Traditional chieftaincies, once looked to for overall dispute resolution as well as overseeing the equal distribution of resources among the community, were perceived to be abusing their positions of power to now control the distribution of compensation and access to employment opportunities. In addition to the erosion of traditional leadership and chieftaincy power struggles, inter-generational conflict worsened as youth began to question the authority of village elders and those viewed as responsible for monopolizing resources. This, in turn, also fueled the rise of cults and vigilante groups, who began to be seen as more representative of particular group interests than traditional leaders (Baker 2012).

From the mid-1990s onward, in this environment of deepening social and ethnic divides and overall inter- and intra-community distrust, violent uprising began to spread across the Niger Delta. As more minority groups began to demand direct access to oil revenues and stake claims to land, gangs comprised of angry and disenfranchised youth also proliferated, and weapons began to flow into the region. Protests and other uprisings were met with increasingly violent tactics by the military, including the formation of a Joint Task Force (JTF), comprised of over 10,000 troops from the various branches of the Nigerian military, in 2003. During that year, in an effort known as "Operation Restore Hope," the JTF launched a campaign meant to quell the increasing restiveness in the Delta, which by now included direct attacks on oil production facilities and pipelines. Kidnappings and abductions were also on the rise as a profitable enterprise, particularly as oil companies often proved willing to pay high ransoms for the return of expatriate executives and other employees. The operation, however, was reported to have resulted in the deaths of hundreds of civilians, as well as the widespread destruction of villages, farmland, and fisheries. Widely condemned by international human rights monitoring organizations, such as Amnesty International, as yet another example of the lack of training and indiscriminate use of force against civilians by the military and police, the campaign further enflamed an already hostile relationship between the federal government and local communities.[1]

It is useful to note that during the above time period, most violent activity was centered in Rivers and Delta states, although there were periodic episodes of violence Bayelsa state, which itself would later also become engulfed in conflict. Rivers was the epicenter of a power struggle between the leaders of two militant groups with deep political and criminal affiliations, known as the Niger Delta Vigilantes (NDV), with factions of the Icelander and Vikings mainly comprising their membership, and the Niger Delta's People Volunteers Force (NDPVF), with members of the KKK, the Greenlanders and Bush Boys (other criminal groups), and independent commanders forming its ranks. The NDV, led by Ateke Tom, was founded in 2003 and initially based in Okrika, Rivers State. Known mainly for its oil bunkering operations and the sabotage of pipelines, the NDV also became renowned for sophisticated piracy attacks in 2007 (Asuni 2009).

The NDPVF, initially the militant wing of the Ijaw Youth Council and sometimes known as the Egbesu Boys until 2003, was led by Alhaji Mujahid Asare-Dokubo, known primarily as Asare. While Ateke Tom was often characterized as little more than a criminal thug, Asare and his group were known for a strongly ideological platform, mainly characterizing themselves as freedom fighters and, in a move that led to an unsuccessful intervention by the government, threatening war on the oil industry in late 2004. Following an unsuccessful October 2004 attempt by the government to disarm and disband both groups by negotiating a peace agreement with Asare, Ateke and their followers, further fragmentation within each group occurred when this agreement ultimately fell apart. Both leaders began to also be seen as disloyal by some of their followers to the original movement at best and, at worst, tools of the government who could be manipulated for cash or other favors (Ibid).

---

[1] See Baker and Francis.

Meanwhile, in Warri, the capital of Delta state, a history of simmering inter-ethnic tensions surrounding access and control of resources had erupted in violence in 1997, 1999 and 2003. Known as the "Warri Crisis" or the "Warri Wars," these episodic waves of violence between the three main ethnic groups, the Ijaw, Itsekiri, and the Urhobo, resulted in hundreds of civilian deaths and the deep entrenchment of rival ethnic identities. These crises also led to the formation of one of the most significant armed groups in the Niger Delta, known as the Federation of Niger Delta Ijaw Communities, (FNDIC). Formed around the perception that the government and oil companies routinely favored the Itsekiris for employment and other opportunities, in part based on a colonial legacy of preferential treatment, the FNDIC quickly became the main platform for greater Ijaw autonomy and rights. Led by Chief Oboko Bello, it was the FNDIC's mobilization officer, known as Tom Polo, who would go on to become Delta State's dominant militia leader and the *de facto* spiritual leader of the movement. He was also the driving force behind the militarization of the creeks, where local youths were recruited to the movement and given arms and training, as well as illegal employment in oil bunkering operations. Polo had begun his career arranging legitimate labor contracts for the main oil companies, and later also providing protection services, and quickly amassed a significant amount of wealth which positioned him as the head of a large patronage network in Warri. It was estimated that by early 2005, Polo had created a sophisticated and highly militarized operation in the creeks, which housed and employed up to 3,000 youths, recruited to the Ijaw struggle.[2]

The Rise of MEND: During this time period, other prominent figures began to emerge as either militant leaders or heads of other splinter organizations. In Bayelsa state, the rise of Victor Ben (Boyloaf) and Henry Okah, following battles with the NDPVF and the support of young militants returning from training in Port Harcourt, was notable. Bayelsa, which never reached the levels of politically-motivated violence during this time period as Delta and Rivers, nevertheless remained a hotbed of Ijaw nationalism which was radicalized following the arrest of Chief Alamieyeseigha, governor of Bayelsa state, on charges of corruption. Asare and the leader of the KKK, known as Olo, were also arrested shortly thereafter while Farah Dagogo, formerly Ateke's second-in-command, and Boyloaf, relocated to Delta state. Following the arrests and the relocation of the two prominent militant leaders to Delta State, and in collaboration with Tom Polo and other senior militants, a new group was formed, known as the Movement for the Emancipation of the Niger Delta (MEND). Included under the MEND umbrella were also members of the cult groups KKK and the Greenlanders. MEND also merged the various oil bunkering syndicates and gave the group much greater financial and logistical power to launch sophisticated attacks on key government and oil facilities (Hanson 2007).

MEND conducted its first attack in December 2005 and although it was repelled, the kidnapping of four foreign oil workers just a few weeks later gave the group the opportunity to articulate its demands to the local and federal government. These demands included not only ransom payments for the hostages, but also compensation

---

[2] See Francis and Asuni.

for environmental degradation in the region and the release of Asare and Alamieyeseigha. Later, MEND would also call for the Niger Delta region to receive 50 % compensation from oil revenues as well as increased political participation, involvement in the oil and gas sector, and a host of socioeconomic development projects. In addition to the articulation of political and social justice demands, MEND also proved to be a sophisticated and formidable enemy, able to reach offshore oil platforms and disrupt oil production. In addition, incidents of kidnappings of foreign oil workers for sizable ransoms were on the rise. In response to MEND attacks, the government often employed military bombardments, razing towns and further radicalizing the local population.[3]

From 2006 through early 2009, MEND's attacks grew more frequent and its ability to cause significant revenue loss through its operations, including sabotaging oil pipelines and attacking offshore platforms, more pronounced. For example, in June 2008, MEND attacked an offshore Shell-operated oil platform, causing an immediate shut down of oil production by an estimated 10 %. This particular attack was significant as it demonstrated the group's ability to mount a complex operation nearly 75 miles offshore. In September of the same year, the group declared an "oil war" with the goal of shutting down all oil operations in Rivers state and attacking government soldiers guarding the pipelines. In addition to attacking and sabotaging the oil and gas industry, MEND also became known for relatively sophisticated kidnapping and ransom operations, although this often caused internal discord as some of MEND's senior leaders either disapproved of kidnapping outright or felt cheated by the split of profits from the ransom. Overall the group—through sabotage, kidnappings, bombings, attacks and oil theft—was able to slow down oil production by up to 40 % on a regular basis, as well as turn much of the Niger Delta into an active combat zone.

Finally, in response to what appeared to be a never-ending cycle of attacks and stoppages in oil production, the government launched a JTF-led military operation against the group across the western Niger Delta in May of 2009. Heavy aerial bombardment as well as ground operations razed villages and caused large-scale destruction to farmland and fishing operations. It also created a huge internal displacement crisis, as thousands of civilians fled the area prior to or in the aftermath of the operation. According to various reports by human rights organizations, the indiscriminate and overwhelming use of force during the operation also resulted in the death of hundreds of villagers, in addition to causing grievous injury to thousands more.

Given the apparently highly organized tactics of the group, as evidenced above, it is important to also note that MEND was always more of an umbrella organization for various armed groups and cults, rather than a static entity. For instance, in the Niger Delta, there emerged three branches of MEND, each with distinct agendas and leadership. In Delta State, it became known as Western MEND and fell under the leadership of Tom Polo. In Rivers, Eastern MEND was formed following a split between Polo and two senior leaders, Farah and Boyloaf, with Farah soon partnering with the head of the Outlaws criminal gang, Soboma George. In Bayelsa, where Boyloaf eventually

---

[3] See Francis and Hanson.

returned, the group was known as Central MEND, and thought to be most closely associated with Henry Okah. Each group engaged in kidnapping and ransom operations, although Eastern MEND, in Rivers, became most widely known as mainly a kidnapping franchise. Western MEND, while also engaging in a wider array of criminal enterprise including the abduction of oil workers, still maintained a charged political platform, however, committed to the self-determination of the Ijaws. Central MEND, through Boyloaf's affiliation with Okah, was often considered the "main MEND" group by outsiders, as this faction continued to release statements claiming to speak on the behalf of the group as a whole, although this was vigorously disputed by the other two factions. Overall, like other militias and armed groups in the region, shifting allegiances and competing social and illicit economic agendas continued, as in the past, to muddy attempts to negotiate and disrupt militant activities.

The 2009 Amnesty: Following the October 2009 JTG assault on the western Delta, which initially targeted Tom Polo's territory and forces but then spread to other MEND bases and commanders, the late President Yar'Adua seized upon an opportunity to negotiate an end to the siege and the introduction of an Amnesty program. Although the President had tried, upon initially assuming office in 2007, to achieve the surrender and disarmament of MEND leaders and their factions through a Presidential Peace Initiative, it was not until the month-long assault on various MEND leaders and their factions that an agreement was finally reached. The Amnesty program, which was part of the agreement, not only offered a full pardon to militant commanders, but also a comprehensive disarmament and reintegration program to combatants. The Amnesty program also put forward a monetary compensation program and skills training program for former combatants, as well as programs for social and community reintegration (Francis 2011).

The Amnesty program, while largely considered successful in restoring a semblance of calm and stability to the Niger Delta region, is not without its detractors. For some, it is viewed merely as a large-scale "combatant buy off" program where senior level commanders have essentially become highly paid entrepreneurs, who are still engaged in illicit activity while drawing large financial compensation packages to refrain from fomenting renewed conflict. In addition, the Amnesty program has also been the source of tension within and among various militant groups and their members, with some drawing larger compensation and "gifts" than others as well as appearing to curry favoritism from the local and federal government. The Amnesty program also rests on the goodwill of the government to continue the payouts and support provided to former combatants. President Goodluck Johnathan, upon assuming office in May 2010, continued the program as inherited from his predecessor although it is scheduled to end in 2015 (Ubhenin 2013).

Despite criticisms of the program put forth both inside of Nigeria and internationally, its gains have nevertheless been notable. As detailed in the data and analyses of Bayelsa, Delta and Rivers states, there has been a sharp decline in kidnappings, abductions, and levels of violent conflict from 2009 to 2013. Attacks on oil infrastructure have also decreased, and productivity has returned to normal or has exceeded normal outputs. The first phase of the Amnesty Program, which included disarmament and demobilization, has been judged relatively successful although by no means completely inclusive. As of late 2012, official reports claim that nearly 27,000 former militants have been disarmed and demobilized. The second phase of the program, with a focus on reintegration of former combatants into communities through the provision of educational and vocational programs, has been deemed less successful as many graduates of these initiatives have returned to the Delta and not found the viable job opportunities promised. The third phase of the program, which includes large infrastructure development plans, has also been rolled out in fits and starts and has not necessarily always provided easy avenues of employment for former combatants (Francis P.). Finally, as data from the three main states also indicates below, militant groups such as MEND are still active, even if their leadership has nominally disavowed violence. There are many factors that have compelled renewed attacks and violence, including the still significant prevalence of arms in the region as well as disgruntled militants who were either left out of the Amnesty program or have found criminal enterprise to still be far more widely available and lucrative than the formal employment market.

While the threat of a large scale return to the creeks and the militancy of the years prior to 2009 poses a threat to nearly every sector, as well as to society as a whole, the politics of the Niger Delta have proven anything but rational in the past, thus presenting no promises for the future. While the Amnesty program has managed to thus far discourage a full-scale return to the militancy of the past, many complain that not enough has been done to address the fundamental conflict drivers that fueled the Niger Delta insurgency. These include a basic lack of access to public services like education and healthcare as well as the ongoing environmental degradation that affects livelihoods and overall health. Human rights abuses, committed by all parties to the conflict, have also not been sufficiently addressed, and tensions continue to bubble beneath the surface. Political factionalization, particularly during election time, normally leads to sporadic but deadly outburst of violence, as noted in the data for the individual states below. Both historical and recent group grievances, based on a multitude of factors, are often manipulated by political figures to stoke the flames of inter-tribal and inter-ethnic rivalries that keeps communities and states tense and distrustful. Thus, while the Amnesty program has achieved many of the short-term objectives of stopping what appeared to be a never-ending cycle of insecurity and violence in the Niger Delta, it remains to be seen whether in the longer term these gains will be proven to be sustainable (Fig. 2.1).

2.1 Abia State

Incidents Per Capita Rank 12/37; Fatalities Per Capita Rank 25/37 (Fig. 2.2).

Abia State has an estimated population of 2.4 million people, predominantly of Igbo origin. Comparatively, it has not experienced the levels of violence and insecurity that other states in the Niger Delta have over the time period analyzed, although there was a sharp uptick in violence in 2010 associated with a

Timeline

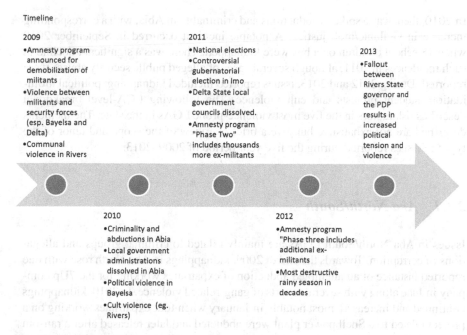

2009
- Amnesty program announced for demobilization of militants
- Violence between militants and security forces (esp. Bayelsa and Delta)
- Communal violence in Rivers

2011
- National elections
- Controversial gubernatorial election in Imo
- Delta State local government councils dissolved.
- Amnesty program "Phase Two" includes thousands more ex-militants

2013
- Fallout between Rivers State governor and the PDP results in increased political tension and violence

2010
- Criminality and abductions in Abia
- Local government administrations dissolved in Abia
- Political violence in Bayelsa
- Cult violence (eg. Rivers)

2012
- Amnesty program "Phase three includes additional ex-militants
- Most destructive rainy season in decades

Fig. 2.1 Timeline of Delta state

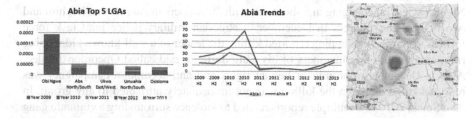

Fig. 2.2 Bar Chart shows annual per capita incidents for Top LGAs; Line Graph shows trend in # of incidents and fatalities in the state; Heat Map shows hotspots of violence from 2009 to 2013—Nigeria Watch data formatted and uploaded to Peace Map (www.p4p-n)

surge in kidnappings. Abia produces about 27 % of Nigeria's crude oil and a significant amount of its natural gas. It is also rich in yam, maize, rice, potatoes, and cashews (www.abiastate.gov.ng).

Theodore Oriji, PDP (People's Democratic Party), was re-elected as governor of Abia state in 2011. In 2013 his government provided buses to transport local traders to and from Umuahia market after the market was relocated as part of a city expansion scheme. Since the dissolution of the local government administrations in January 2010, there have been no Local Government Area (LGA) level elections, as of December 2013. From 2009 throughout 2010, the main drivers of insecurity in Abia were kidnappings and abductions, with some gang-related and police violence.

In 2010, there was a spike in abductions and criminality in Abia, with a corresponding increase in vigilante/mob justice. A notable incident occurred in September 2010 when 18 school children on a bus were kidnapped. There was a significant decline in such incidences in 2011, although several cases of alleged public security abuses were reported. During 2012 and 2013, issues reported included kidnapping, political intimidation, student protests and cult violence. The following LGA-level breakdown describes risk factors in the five most violence-prone LGAs in the state. The incidents described are not exhaustive but give a brief overview of the scope and tenor of the types of issues reported during the five year period of 2009–2013.

2.1.1   Aba North/South

Issues in Aba North/South LGAs are mainly related to criminal groups and allegations of corruption. Towards the end of 2009, kidnappings in Aba North rose with one reported instance of an attempted abduction of expatriates working for the 7Up company in June along with several reports of gang-related violence. In 2010, kidnappings continued and increased, most notably in January when four expatriates working on a project related to a Shell power plant were abducted and later released after a ransom was paid. The year continued to see multiple instances of kidnappings and abductions, some attempted and others successful. While the focus of the kidnappings appeared to be foreigners and expatriate workers, locals were also targeted, with reports of a Nigerian General engaging in a shoot-out with kidnappers in August that left him and two of his aides injured. In December, a Nigerian military-police joint task force formed to combat kidnapping reported killing the leader of a well-known kidnapping ring, along with other members of his gang during a raid called Operation Jubilee. Also in December, a prominent human rights and anti-corruption campaigner was reported killed although the killing was not immediately attributed to any group. In Aba South, 2010 saw multiple reports related to violence surrounding a vigilante gang called the Bakassi Boys, with various reports indicating that human rights abuses of civilians and suspected kidnappers had been perpetrated by the gang. Beginning in September when it was reported that banks and local businesses had been shut down for weeks due to insecurity, reports of professionals fleeing the LGA for other areas began to increase. In 2011, there was a marked decrease in violence in both LGAs, likely as a result of increased police and military activity in the area and a focus on routing out criminal and kidnapping gangs. In 2012–2013, while insecurity remained the most reported on indicator in the two LGAs, incidences of kidnapping and targeted violence dropped while incidences of general insecurity related to petty crime and political corruption remained relatively steady. In July 2012, there was a protest by women's groups about layoffs in various sectors by the state government, while in September and October, there were allegations in local newspapers about the connections between criminal gangs and political parties. Also during this time period there were protests and complaints from various sectors of society about excessive taxation, yet none reportedly turned violent.

2.1.2   Umuahia North/South

Political thuggery, kidnapping, and several cases of cult violence were reported in
Umuahia North/South LGAs throughout the time period examined. From 2009 to
2011, there were several reports of attempted kidnappings, although most were not
successful or were resolved without further violence. In May 2009, it was reported
that police had killed a known kidnapping ring leader after he had abducted and
killed a bank manager. In July and August 2010, there were reports of kidnappings
of journalists, as well as a bus of school children for ransom although no follow-up
reports indicated how they were resolved. In 2011, the main incidents of insecurity
were related to arrests that occurred during protests as well as general criminality,
such as robberies and domestic disputes. In early 2012, the People's Progressive
Alliance headquarters was reportedly attacked and property destroyed by gangs
believed to be connected to opposing political parties. In January 2013, a senior
lawmaker was reportedly kidnapped, while in March of the same year gunmen
reportedly attacked the home of former governor Orji Uzor Kalu.

2.1.3   Obi Ngwa

Obi Ngwa, with a population of less than 200,000 according to a 2006 census, had
the highest number of incidents in Abia during the time period examined, with 35
reported. It was followed only by Aba North/South, which, with a population of
over a half million, reported 28 incidents.

For the most part, Obi Ngwa reported few incidents in 2009, 2011, 2012 and 2013.
Those that were reported mostly concerned domestic violence and petty theft, with the
deaths of three policemen reported from April to September 2009. In 2010, however,
reports of insecurity increased dramatically after multiple kidnappings led to increased
efforts by police to combat the phenomenon which, in turn, led to dozens of police
deaths. Beginning in January 2010, there were monthly reports of police deaths resulting from operations against kidnapping rings. In addition, four journalists were kidnapped in July and a reported 250 million Naira ransom set for their release. This
incident, along with the police slayings and reports of businesses closing or moving
out of Obi Ngwa due to insecurity, prompted the president of the Senate in Abuja to
demand the president call a state of emergency for Abia and surrounding states
plagued by kidnapping. In addition, according to various newspaper reports, he also
advocated tougher tactics by police and joint task force personnel against kidnappers
and armed thugs, declaring that "jungle tactics must be met with jungle tactics."

2.1.4   Ukwa East/West

In the time period examined, most incidents of insecurity that occurred in Ukwa
East/West concerned police violence and kidnapping. In January 2009, it was
reported that JTF personnel trailing a stolen bullion vehicle into a village ended up

destroying property, including homes, and leaving residents homeless. Later that year it was reported that the former Nigerian ambassador to Ukraine and Gabon has been murdered along with his driver and escort in what appeared to be a highway robbery. In 2010, reports of kidnappings dominated, with two German expatriates reportedly kidnapped in April and, in two separate incidents, a pastor and school-children being abducted for ransom. In May, it was reported that kidnappers killed two policemen and a civilian in the abduction of the Chairman of the Abia State Independent Electoral Commission. In addition to the reports of kidnappings, violence related to police and JTF action against suspected kidnappers also accounted for reports of insecurity in 2010, including burning homes of suspects. There were no reports in 2011–2012, although the end of 2013 saw a spike in insecurity surrounding police action against oil bunkering suspects accused of stealing a petroleum tanker.

2.1.5   Osisioma

Violence related to domestic disputes, robberies, and kidnappings characterized the insecurity experienced in Osisioma from 2009 to 2013. As in other LGAs in Abia, violence between police, the JTF, and armed robbers and kidnappers increased in 2010, with two reported police deaths in late August. In April, a protest over the construction of a market reportedly left at least two dead and several injured in the aftermath, as well as causing property damage. 2011–2012 was relatively quiet although insecurity increased again in 2013 with several instances of kidnappings, in two cases of infants suspected to be sold to "baby factories." There were also reports of domestic violence during the year.

2.2   Bayesla State

Incidents Per Capita Rank 6/37; Fatalities Per Capita Rank 8/37 (Fig. 2.3).

With 1.7 million people, Bayelsa is one of the smallest states in the country, by population. Most are of Ijaw descent. Bayelsa produces between 30 and 40 % of Nigeria's oil and gas. In addition to the petroleum sector, the state also has an extensive commercial fishing industry and produces oil palm, raffia palm, rubber, and coconut.

In February 2012, Henry Dickson (PDP) was elected as governor after a period of uncertainty in the wake of Governor Timipre Sylva's termination in January 2012.

In the first half of 2013, Bayelsa fell below the Niger Delta regional average in reported incidents of insecurity per capita for the first time since 2010. On June 15, 2013, thousands of cult group members met with political leaders as part of Gov. Dickson's Restoration Agenda, to renounce criminality and work towards peace.

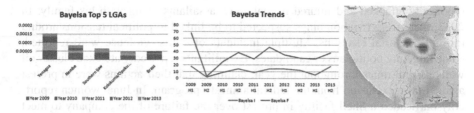

Fig. 2.3 Bar Chart shows annual per capita incidents for Top LGAs; Line Graph shows trend in # of incidents and fatalities in the state; Heat Map shows hotspots of violence from 2009 to 2013— Nigeria Watch data formatted and uploaded to Peace Map (www.p4p-nigerdelta.org/peace-building-map)

Since 2009, incidents of insecurity in Bayelsa included cult violence, abductions, and attacks on energy infrastructure. Conflict factors were mainly reported around the capital of Yenagoa. The following LGA-level breakdown describes risk factors in the five most violence-prone LGAs in the state. The incidents described are not exhaustive but give a brief overview of the scope and tenor of the types of issues reported in the data during the five year period of 2009–2013.

2.2.1 Yenagoa

From 2009 to 2010, reports of insecurity increased markedly in Bayelsa, as reports about gang and cult-related violence began to take on a political dimension in 2010. While the reports in 2009 mainly related to violence between police and JTF and reported cultist groups and armed gangs, beginning in April of 2010, reports of attacks targeting political aspirants or current leaders began to gain prominence. In early April, the Bayelsa Youth Association took to the streets to reportedly protest the nomination of the leader of the local Executive Council while in May a car loaded with explosive detonated outside the home of the Deputy State Governor in what police characterized as an attempted assassination. In June, the home of the Deputy Speaker of the House of Bayelsa state was attacked and another assassination attempt reportedly made against the Deputy State Governor, who was facing an impeachment trial. He was attacked again in July and his bodyguard was reportedly killed. In December, a bomb exploded during a PDP political rally, although there were no reports of casualties. Other issues reported in 2010 concerned protests by militants regarding a lack of payment as promised under the Amnesty Law and battles between police and suspected militants thought to be stockpiling explosives. In addition, general reports of insecurity surrounding robberies and family disputes rose in 2010. In 2011, frequent clashes between police and reported ex militants and members of cult gangs continued. Also, in June, heavy flooding reportedly displaced 600 families in Yenagoa while washing out bridges and roads. In August, clashes between two cult groups thought to be associated with political parties reportedly left 15 people dead while at year's end a local political candidate was reportedly attacked and injured by unknown assailants along with his family. In 2012, there were two bombings suspected to be linked to political tensions around the gubernatorial elections. Multiple incidents of cultist violence were reported, some targeting politicians or candidates. Ex-militants reportedly attacked cluster oil wells several times throughout the year, claiming that their actions were in protest against their exclusion from the federal amnesty program. In June, women reportedly barricaded a Shell facility in protest over the failure of the company to meet community obligations while causing environmental and social degradation. In November 2012, hundreds reportedly protested the non-payment of allowances provided by the amnesty program over a six-month time period. Also in November, hundreds of flood victims reportedly protested against government action to remove them from relief camps. There were multiple reports of police violence against civilians throughout the year, mostly during clashes with suspected armed robbers and kidnappers. In 2013, similar incidents of violence surrounding the exclusion or the perceived inadequate inclusion of ex-militants in the amnesty program characterized the first half of the year. In February, for example, it was reported that up to 200 former militants rioted and destroyed property when they learned that they would not be included in the third phase of the amnesty program. In August, *Vanguard* News reported that 12 suspected pirates killed by the Nigerian Navy were actually members of a defunct militant group refusing to turn over arms or join the amnesty program. November of 2013 was characterized by violence reportedly stemming from continued political tensions surrounding the Ijaw National Congress Elections in October. Incidents included the reported targeting and kidnapping of youth group members and political allies supportive of each candidate. Later in November, communal violence reportedly killed up to five people in a clash between Agudama-Epie and Akenfa communities over farmland.

2.2.2 Nembe

From 2009 to 2011, most reported incidents of insecurity were attributed to clashes between MEND and security forces, particularly around attacks on oil infrastructure. Throughout 2009, there were multiple reports of oil pipe lines being blown up and, in one case, an oil platform temporarily being taken over in March before the suspects were reportedly killed by the police. In August 2009, a prominent leader of MEND was reportedly killed by police although later details appeared to indicate that he was on his way to a location to surrender weapons, which was never verified. In late 2011, most reports concerned clashes on water between suspected sea pirates and marine police. In 2012, incidents reported in Nembe included an alleged attack by MEND on marine policemen in 2012. Data from CSS/ETH Zurich and the Council on Foreign relations reported an April 2013 attack by MEND on an oil well which was said to have caused a spill and created an environmental emergency. In the second half of 2013, however, violence spiked surrounding attacks by reported pirates on passenger boats as well as military police and Navy formations. It was reported that as of late 2013, up to twenty civilians had been killed by pirate activity or from being caught in the crossfire between police and suspected pirates. This number could not be independently verified, although multiple news sources did corroborate the rise in violent deaths and general insecurity surrounding suspected pirate activity and police counter-actions.

2.2.3 Ijaw South

From 2009 to 2010, reports of violence and insecurity were mostly attributed to attacks on oil infrastructure and resulting police clashes with MEND. In February 2010, it was reported that oil spillages had already accounted for the deaths of at least 30 children that year, prompting a local MEND leader to declare the Amnesty Law and its provisions a failure. Later that year, there were reports of police raiding suspected militant camps in search of weapons in November while a feud between the Governor and Deputy Governor continued, reportedly adding to inter-communal tensions between supporters of each individual. Reports of violence decreased in 2011 with a landslide capturing the news in September which led to reported pro-tests and accusations by community leaders that oil dredging by foreign companies had led to the disaster. In 2012, reported incidents of insecurity included attacks on energy infrastructure, politically motivated violence, and cult attacks on university students. In January 2012, unidentified gunmen reportedly attacked the home of an ex-militant leader and killed a policeman and young ex-militant. A fight broke out at a political rally in February 2012, resulting in at least one death. In March and April 2012, there were reported attacks on oil pipelines, the first such reports in over a year. In July, gunmen reportedly attacked a boat belonging to an oil com-pany, killing at least three and injuring others. Throughout the year cultists report-edly killed Niger Delta University students in several incidents. Flooding was a problem in October 2012. In 2013, there were reported clashes between members of the Joint Task Force (JTF) and militants in the Azuzuama area, killing several and displacing local residents. The Movement for the Emancipation of the Niger Delta (MEND) also claimed to have killed 15 officers in a boat attack. In early May 2013, a shooting by suspected renegade militants left five ex-militants dead. In the second half of 2013, reports of police action leading to the deaths of suspected pirates, as well as police deaths, were reported in May, July and October, although the estimated numbers of individuals killed varied. Other incidents reported in 2013 included deaths related to oil bunkering as well as a few instances of kidnap-ping of local oil dredgers.

2.2.4 Kolokuma/Opokuma

In 2009 and 2010, there were fewer incidents reported of general insecurity, although a flood in October of 2009 reportedly resulted in the displacement of 5,000 people. In November 2010, it was reported that a gang bombed the home of the Presidential

Adviser on Niger Delta Affairs and Coordinator of the Federal Government amnesty, killing a policeman. No further details, or claims of responsibility from any group, emerged in the aftermath, however. In January 2011, clashes between rival gangs who reportedly received support from political groups left six dead, although there were no other incidents reported that year. Incidents reported in Kolokuma/ Opokuma in 2012 mainly related to the devastation caused by flooding. In March 2012, however, it was reported that a "general" from a local militia had attacked an oil flow station in protest of the lack of development in the Niger Delta. In mid- to late-2013, violence related to domestic disputes were the main reported incidents of insecurity, although early in the year flooding devastation still lingered.

2.2.5   Brass

With its extensive shoreline, Brass experiences frequent attacks on nearby vessels, many of which are associated with the oil industry. In 2009, there were multiple reports of attacks by suspected MEND militants as well as gang members on oil infrastructure that led to battles with police and the JTF. In November 2010, it was reported that a prominent militant leader was killed with a machete although he had reportedly accepted amnesty. In 2011, attacks on oil platforms and other infrastructure remained prominent, with a reported kidnapping incident involving at least three foreign nationals occurring in November. It was reported that all were later released unharmed. Also that year, in April, disputes between rivals from the PDP and LP political parties also were reported, with accusations of the theft of electoral materials surfacing. From 2012 to 2013, data from CSS/ETH Zurich points to over a dozen such incidents which resulted in kidnappings as well as the loss of property and lives. In June 2012, a JTF/Navy patrol reportedly killed six pirates on a vessel transporting 8.5 L of stolen crude. Three naval officers were also killed in the fight which later became a source of tension in the community as the Independent newspaper reported that the suspected pirates were actually indigenes returning from a funeral and were targeted by the patrol. A community petition was then reported to have been circulated and sent to the Chief Army Staff protesting the killing and requesting an inquiry into extrajudicial violence undertaken by such patrols against civilians. In November of 2013, it was reported that a group of seven gunmen kidnapped the father of Bayelsa State Commissioner for Tourism, although no specific reasons were given. Other reported issues during this time period included political tensions shortly after the election of Governor Dickson. In particular, five LGA chairmen were removed from office for financial recklessness shortly after the election in October 2012. Although they denied the charges, accusations were made that Dickson was targeting certain individuals in order to marginalize them from politics. These accusations, and other related claims, continued throughout 2012–2013 although were steadily on the decrease by the end of 2013.

2.2.6 Other LGAs

2.2.6.1 Sagbama

In addition to general insecurity, 2009–2010 was characterized by reported clashes between suspected militants and the JTF, including a botched kidnapping attempt of foreign nationals in April 2010 that led to the deaths of four kidnappers. In January 2011, one person was reported killed in a clash between individuals suspected to be mercenaries who had been hired to oversee voter registration. In October 2012, massive flooding hit Sagbama causing food scarcity and epidemics, as Nigeria experienced the most severe rainy season in decades. That same month, political tension was reported when the LGA council chairman was removed for alleged impropriety. In November 2012, there was a reported incident of intra-communal conflict which resulted in the death of about a dozen people after a traditional ruler was removed by the Sylva administration. In February 2013, an incident of piracy was reported in which gunmen allegedly killed several soldiers who were escorting an oil vessel in the creeks. In December there was a report of possible inter-communal tension when two men were killed after a feud with herdsmen over cattle blocking a main roadway although it appeared to be an isolated incident. Tension within the Ijaw Youth Council was also reported at intermittent times throughout the year.

2.2.6.2 Ogbia

From 2009 to 2013, reported issues in Ogbia included sexual violence and child abuse. In February 2011, politically motivated violence was reported when gunmen invaded a community and reportedly fired randomly at villagers as Goodluck Jonathan's wife was participating in a nearby ceremony. In November 2011, the deputy chief of a local community was reportedly kidnapped, setting off accusations that the kidnapping was politically motivated, although no further evidence emerged. In 2012, as in Sagama, political tensions were on the rise following the election of Governor Dickson, although they largely appeared to have subsided by the end of the year. Flooding reportedly submerged communities in October 2012 and led to displacement and some food scarcity issues. In January 2013, gunmen reportedly invaded the country home of the local Council Chairman and abducted his parents. By year's end it was not clear whether they had been released but the story drew wide media attention as a testament to Bayelsa's ongoing epidemic of kidnapping and abductions.

2.3 Akwa Ibom State

Incidents Per Capita Rank 21/37; Fatalities Per Capita Rank 23/37 (Fig. 2.4).

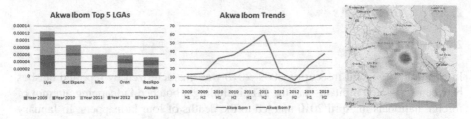

Fig. 2.4 Bar Chart shows annual per capita Incidents for Top LGAs; Line Graph shows trend in # of incidents and fatalities in the state; Heat Map shows hotspots of violence from 2009 to 2013—Nigeria Watch data formatted and uploaded to Peace Map (www.p4p-nigerdelta.org/peace-building-map)

Akwa Ibom has a population of about 3.9 million people according to the 2006 census. Predominantly inhabited by the Ibibio people, the state is also home to Annang, Oron, Obolo and Eket communities. Endowed with large deposits of crude oil, condensate and gas, Akwa Ibom is among the largest petroleum producers in Nigeria. Agriculture also constitutes an important income-generating activity in the state, particularly the farming of palm produce, rubber, cocoa, rice, cassava, yam, plantain, banana, maize, and timber (www.akwaibomstate.gov.ng).

Violence in the state escalated during the gubernatorial elections of 2011. But after the re-election of Governor Godswill Obot Akpabio (PDP), however, the per capita level of violence dropped significantly. In 2013 violence trended upward, with incidents reported around the capital city of Uyo, the town of Ikot Ekpene, and the coastal Local Government Areas (LGAs) to the south, including issues of land conflict, political tensions, protests, and abductions. From 2012 to 2013, however, Akwa Ibom was the least violent state in the Niger Delta region as measured by reported incidents per capita. The following LGA-level breakdown describes risk factors in the five most violence-prone LGAs in the state. The incidents described are not exhaustive but give a brief overview of the scope and tenor of the types of issues reported in the data during the five year period of 2009–2013.

2.3.1 Uyo

From 2009 to 2010, there were regular reports of insecurity, mostly stemming from robberies, domestic violence and suspected cult abductions. There were two incidents of attacks on an oil vessel and a platform in May and July of 2009, although there were no further details reported. Also during this time period, multiple instances of kidnappings for ransom were reported, with children and, in one instance, several Lebanese expatriates as targets. In almost all incidents, ransom was paid and/or the victim was released or rescued without further injury. In 2011, there were several incidents of violence surrounding the election as well as the continuation of kidnappings, robberies, and cult-related violence. In March, it was reported that a student march on Ibom Hall following a government speech regarding empowering students for the elections resulted in a stampede which reportedly left several dead. There were also several instances of what appeared to be politically-motivated kidnappings and killings. In particular, in April it was reported that a popular journalist had been killed for voicing anti-government sentiments. Then, in May, an Action Congress of Nigeria (ACN) chieftain was reportedly murdered by assassins although no specific motive was reported. Also in May, the ACN challenged the election of Governor Godswill Obot Akpabio, claiming vote rigging and fraud. Following those allegations, it was reported in July that a member of the Independent National Election Commission went into hiding following death threats amidst claims that he had played a role in ballot rigging. By mid-year, however, most election-related violence had subsided, although instances of general insecurity persisted. In September and October, there were reported attacks on Exxon oil platforms by pirates and robbers, although neither resulted in any fatalities. In 2012–2013, there were a series of killings reported, including that of a former government official in October 2012. A violent clash among three rival cult groups was also reported in October 2013, killing seven. Additionally, in 2013 there were multiple violent protests, including one in April against a company's employment policies. The University of Uyo (UNIUYO) was forced to briefly shut down in June after a protest turned violent, destroying school property and killing several students.

2.3.2 Ikot Ekpene

While there were no reported incidents in 2009, insecurity spiked in Ikot Ekepne in the first half of 2010. From January until May of 2010, there were four reports of insecurity, mainly concerning domestic violence and cultism. In May, however, there was a reported foiled robbery attempt resulting in the death of at least two alleged criminals. In 2011, insecurity characterized by political violence led to the deaths of at least 12 people while criminality and a student protest that turned violent led to the death of at least two more people. In March, a clash between supporters of rival PDP and ACN candidates for governor turned bloody when soldiers were called in to quell the violence. Earlier in the year, gunmen reportedly killed a man who was vying for a vacant position as a traditional ruler in the state and had allegedly quarreled with the other contender. Apart from some alleged irregularities reported during the April 2012 PDP primaries, Ikot Epkene LGA had few reported deadly incidents until the latter half of 2013, when there was reported to be a failed rescue operation of an abducted politician in October and a clash between youth settlers and herdsmen over land and grazing rights, allegedly claiming one life in December.

2.3.3 Mbo

While there were no reported incidents in 2009, an oil spill in May 2010 led to protests by residents at the start of the month as well as women's groups later in the same month. In 2011, however, attacks on offshore oil platforms and vessels increased, beginning with an April attack on a supply vessel that resulted in the abduction of the boat's captain and stolen valuables. Similar attacks were attempted in September and November of the same year although, in both instances, they were unsuccessful and no one was abducted or killed. An incident of communal violence was also reported in November, leading to the alleged death of two individuals. Mbo LGA was affected by intra and inter-communal tensions in 2012–2013. In May 2012, two communities clashed, reportedly killing one. In January 2013, seven reportedly died in a separate clash over farming land. In March 2013, there was a reported clash in Unyenge community. In November 2013, two women were killed in a renewed clash among communities. Mbo LGA has also been affected by piracy including reported incidents in February 2012 and August 2013, leading the government to increase its naval presence in the coastal waters. In July and September 2013, it was reported that there were violent protests over the lack of amenities and compensation by companies operating in the area.

2.3.4   Oron

Among LGAs in Akwa Ibom, Oron reportedly has a relatively high concentration of cult groups and cultist activities. While there were no reported incidences of insecurity in 2009, 2010 opened with a bloody clash over communal fishing rights between two communities which allegedly led to the deaths of at least 12 people and the destruction of a reported 40 homes. Later in the year, in October 2012, four people died in a rivalry between the Black Axe and Vikings cult groups. Separately, at least one former official was reportedly killed for political reasons in 2012. In August 2013, there was a gun battle between the navy and a group of pirates, killing six suspected pirates.

2.3.5   Ibesikpo Asutan

From 2009 to 2010, there were three incidents related to insecurity reported in Ibesikpo Asutan. In 2009, the murder of the owner of a filling station was reported initially as an assassination, although no further linkages could be made. In June 2010, a prominent local businessman, and chieftain of the PDP, was reportedly killed while attending church, although no reasons were immediately reported. In October of the same year, another local PDP official working for the former Governor was allegedly murdered, again with no motive being cited. In 2011, violence erupted in April when it was reported that individuals supporting the PDP attempted to steal election materials during the gubernatorial elections. Five people were reported killed during the incident with several others injured, including INEC staff and members of the security force hired in case of violence. From 2012 to 2013, there were two incidents of insecurity reported, leading to the death of two people. The first was a domestic violence issue surrounding the murder of a newborn in February 2012 while in March 2013, a local trader was reportedly murdered by suspected thieves.

2.4   Rivers State

Incidents Per Capita Rank 10/37; Fatalities Per Capita Rank 15/37 (Fig. 2.5).

Among the largest of the oil-producing Nigerian states, Rivers had been at the heart of the Niger Delta militancy until 2009, when violence was at its highest during the period examined. Following the enactment of Amnesty, violence dropped dramatically, although the state is beset with a different array of issues as some former combatants have turned to criminality and uneven economic development continues to pose a challenge to sustainable peace and human security. Since 2010, the number of fatalities associated with conflict risk factors has decreased slightly. Of the 23 Local Government Areas (LGAs) in Rivers State, those with the highest number of reported incidents per capita resulting in fatalities were Port Harcourt, Ahoada East, Eleme, Asari-Toru, Ogba/Egbema/Ndoni, Khana, Ikwerre, Obio/Akpor, and Emohua. Since May 2013, political tensions have been elevated in Rivers State after the disputed Nigerian Governor's Forum election, in which Rivers State Governor Rotimi Amaechi was the incumbent. On July 9, a fight broke out in the Rivers State House of Assembly following a move to impeach the Speaker of the House, who is considered to be a supporter of Governor Amaechi. In mid-July, when the governors from four northern states came to Rivers to visit Amaechi in a purported show of solidarity, protestors demonstrated at the airport. Concerned about the polarization and political maneuvering in the run-up to the 2015 elections, the Niger Delta Civil Society Coalition organized a rally on July 30 to promote freedom of assembly and democracy. Nevertheless, protests both for and against the governor continued throughout the year, usually ending with a non-violent police intervention. The following LGA-level breakdown describes risk factors in the five most violence-prone LGAs in the state. The incidents described are not exhaustive but give a brief overview of the scope and tenor of the types of issues reported in the data during the five year period of 2009–2013.

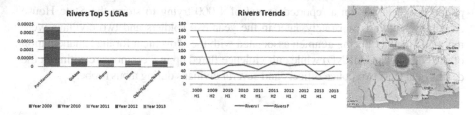

Fig. 2.5 Bar Chart shows annual per capita incidents for Top LGAs; Line Graph shows trend in # of incidents and fatalities in the state; Heat Map shows hotspots of violence from 2009 to 2013—Nigeria Watch data formatted and uploaded to Peace Map (www.p4p-nigerdelta.org/peace-building-map)

2.4.1   Port Harcourt

Port Harcourt, the capital of Rivers State, has the highest population of all the state's LGAs. Given the high density of the urban population, conflict risk issues include robberies, kidnappings, and gang/cult-related violence. As the political capital of the state, protests and demonstrations are common. In 2009 alone, over 50 fatalities were reported, mostly related to attempted kidnapping and rescues. These kidnappings ranged from the abductions of children of prominent community businessmen to the attempted abductions of foreign employers of oil companies. In several cases, police were reportedly murdered attempting to prevent the kidnapping or during a botched rescue attempt. Also during 2009, clashes between the JTF and cult groups also led to numerous fatalities, most notably in February when at least ten members of an alleged cult group drowned after being chased by officers reportedly belonging to the JTF. In October of 2009, a report by Amnesty International condemned the JTF for using disproportionate force against communities after at least ten people were reported killed during a protest over the demolition and relocation of community homes. In 2010, fatalities remained high as most incidents occurred between police and various armed groups, including cult gangs. In March, at least seven police officers were reported killed after a shootout with an armed gang. In August, a well-known militant was reportedly gunned down by a cult gang in an incident that also resulted in the death of at least one bystander and the injury of several others. In 2011, most incidents of insecurity reported were again abductions and kidnappings, robberies, and cult violence. A report in June of 2011 proclaimed it to be one of the deadliest years for police, with at least 25 reportedly gunned down in the line of duty. In February alone, it was reported that two separate attacks on money transportation trucks led to the theft of at least 50 million Naira. The year closed with the reported killing of two thieves dressed in military uniforms who attempted to steal petrol from a local station.

In the first half of 2012, large protests broke out over the government's removal of a fuel subsidy program. Later in the year, there were smaller protests reported over poor delivery of public services. In the first half of 2013, there was increased political tension between supporters and detractors of Governor Amaechi, including a protest where police fired tear gas on a reported crowd of 1,000 trying to storm the State House Assembly building. Separately, later in the year two soldiers and two civilians were reported killed by fleeing gunmen associated with the Movement for the Emancipation of the Niger Delta (MEND) while cult violence flared up in December, resulting in the deaths of at least two police officers and several civilians during a bus robbery.

2.4.2   Gokana

Gokana, with a population of just under 230,000 according to the 2006 census, experienced high levels of insecurity in 2009 and 2010, particularly intra-communal violence with one ethnic group split between two rival cult gangs. In February 2009, two separate attacks, blamed on disagreements between the local Ogani communities and an international oil company, led to at least ten reported deaths and dozens injured. Relations between the local communities and the oil company remained particularly fractious throughout the year, with various groups claiming to either be fighting for or against the interests of the local community or the oil company. In June, it was reported that MEND managed to free a captured foreign oil worker during an operation where the JTF also claimed to assassinate a militant leader. In 2010, two major intra-communal clashes between a community split by loyalties to rival cult groups led to the reported deaths of at least 26 people from mid-October to the end of the year. It was also reported that there was significant property damage, causing temporary relocation of communities. In 2011, violence dropped sharply from the previous two years, with the main incidents reported continuing to be clashes between rival cult gangs, with four reported fatalities. There were no reports in 2012 while the one report in 2013 again concerned the reported kidnapping and murder of at least two individuals from rival cults. The rest of the year remained quiet as far as reported violence.

2.4.3   Khana

In Khana LGA, there were no incidents reported in 2009 and one in 2010; specifically, a youth protest that reportedly led to the death of two police officers when events turned violent. There was an uptick in violence in 2011, with 12 reported fatalities during the year. In April, it was reported that a member of the ACN was killed by thugs purportedly associated with the PDP during the gubernatorial elections. No motive was given for the attack. In June, two youths were reportedly killed by police during a confrontation over land that had been sold by village elders to the military for a cantonment site. In December, robberies led to at least six deaths, with five reported deaths occurring during a botched bank robbery. In 2012, there continued to be communal tensions, particularly around the issue of land. There were also cult clashes and criminality including robbery and kidnapping. Land competition reportedly turned violent in 2012 when two communities in Ogoniland disagreed over whether and how a parcel of land should be developed by the state government as a banana plantation. In 2013, police arrests connected with cult violence predominated in the beginning of the year while a land seizure related to the development of a banana farm sparked tensions in late May.

2.4.4   Eleme

In August of 2010, there was one reported fatality in Eleme after a cleric's family was saved from abduction by a police officer who was killed in the struggle. In 2011, there was also one reported fatality when a human rights activist was allegedly murdered after declaring his intention to run for political office. In March 2012, a solider apparently shot and killed a boy after he reportedly threw a bottle in the direction the soldier's car, resulting in a protest by community youth over the death. In 2013, cult violence was blamed for violence and deaths, most notably in May when two people were reported killed in a clash between two rival groups. Several robberies during the year also resulted in deaths, with a reported three people being killed after an attempted break-in on a residential building.

2.4.5   Ogba/Egbema/Ndoni

In Ogba/Egbema/Ndoni the primary issues reported during this period include flooding, criminality, kidnappings, and gang violence. Between 2009 and 2011, there were at least six fatalities reported, mostly during kidnapping operations which were interrupted by police. In November 2011, however, a youth protest over the death of a local worker at a flow station resulted in the allegedly death of two when soldiers intervened, attempting to reclaim seized weapons. In October 2012, severe floods caused the displacement of entire communities leading to inflation, starvation and serious difficulties with resettlement. In May 2013, unknown gunman reportedly tortured and killed an aid to the former PDP chairman Chief Godspower Ake.

2.5   Delta State

Incidents Per Capita Rank 2/37; Fatalities Per Capita Rank 3/37 (Fig. 2.6).

Delta is the second most populous state in the Niger Delta, with an estimated 4.1 million people. The state produces about 35 % of Nigeria's crude oil and a considerable amount of its natural gas. It is also rich in root and tuber crops, such as potatoes, yams, cassava, and coco yams. Delta has a legacy of ethnic and political tensions which flared in the late 1990s and again in 2003. The 2009 Amnesty Program was instrumental in reducing violence and fatalities associated with militancy. In 2010, however, there was a spike in insurgency/counter-insurgency activity in the Burutu Local Government Area (LGA) in December. In June 2013, it was announced that LGA-level elections would be held after a delay lasting over two years since May 2011 but as of January 2014, the elections had not yet taken place. During 2012 and 2013, reported incidents included gang violence, criminality, and vigilante/mob justice. There were a number of abductions, some of which targeted political figures, their family members, or oil workers. There were several reports of alleged abuses by public security forces, which sometimes provoked mob violence and protest. The following LGA-level breakdown describes risk factors in the five most violence-prone LGAs in the state. The incidents described are not exhaustive but give a brief overview of the scope and tenor of the types of issues reported in the data during the five year period of 2009–2013.

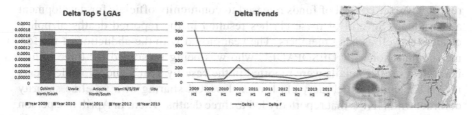

Fig. 2.6 Bar Chart shows annual per capita incidents for Top LGAs; Line Graph shows trend in # of incidents and fatalities in the state; Heat Map shows hotspots of violence from 2009 to 2013— Nigeria Watch data formatted and uploaded to Peace Map (www.p4p-nigerdelta.org/ peace-building-map)

2.5.1 Oshimili North/South

From 2009 to 2010, most reported incidents of insecurity in Oshimili N/S concerned kidnappings, domestic disputes, robberies, and abductions. There were also deaths reported related to ritual sacrifices and cult violence. In October 2010, a local businessman and oil marketer was reportedly found dead in what was believed to be his abductor's base of operations after a 70 million Naira bribe was unable to be secured. In 2011, most incidents reported also concerned domestic violence, road accidents resulting in mob attacks, and botched robberies. In 2012, there was some reported tension between settler and indigenous communities. In one reported case there was a clash between Hausa traders and Igbo youth in February. In another case, two Fulani herdsmen were reportedly lynched. There were also kidnappings, mob violence, and cult attacks reported throughout the year. In 2013, there were protests surrounding the allocation of traditional tribal land for use by an oil flow station. Separately, a local chief was reportedly abducted for ransom and later murdered by his kidnappers. Armed robberies and general criminality continued throughout the year.

2.5.2 Uvwie

In 2009, the main incidents of insecurity reported in Uvwie were related to attempted oil thefts that resulted in police and JTF action as well as a major dispute in October between rival gangs over the reported leadership of a criminal syndicate. In June 2010, a protest against the Delta State Oil Producing Areas Development Commission (DESOPADEC) was held by local residents who demanded that 50 % of oil proceeds be returned to the community. In November a CAN political candidate reportedly narrowly escaped being killed when a bomb detonated outside his residence, killing his elderly mother and young niece. Throughout the year violence and general insecurity rose as rival gangs clashed, reportedly over control of funds released to community officials for development and training. Several of these clashes resulted in the reported deaths of police officers as well as gang members. The year closed with the abduction of the former commissioner of DESOPADEC who was being ransomed for three billion Naira. No further details were reported in the case by year's end. Notable events in 2011 included a February clash over the local sharing of a pipeline security contract with NNPC that reportedly led to three deaths. Then, in April, one person reportedly died following the conclusion of governorship and House of Assembly elections that caused a community disagreement over the results. A resulting protest reportedly led to the deaths of two more individuals. In the first half of 2012, there were incidents of clashes between rival gangs as well as a May youth protest regarding the slow pace of the development of a road between Port Harcourt and Warri which included local infrastructure and development projects. In September, it was reported that youths attacked a police station in violent protest over the killing of an ex-militant Ijaw commander. Also that month, there was an ultimatum issued by a local community to a foreign oil company demanding that they vacate community land following alleged maltreatment of community leaders and general neglect of the land and local population. In July 2013, it was reported that environmental pollution caused by the activities of an unnamed construction company allegedly led to the death of three persons. In December, intra-communal violence over land issues and community leadership resulted in a violent clash reportedly led by local youths. It was later reported that residents fled following the outbreak of violence and economic activity was halted, although there were no subsequent reports of deaths or property destruction.

2.5.3 Aniocha North/South

As in other parts of Delta State, much of the violence in the reported time period in Aniocha North and South was associated with kidnappings, domestic violence, and abductions. From 2009 to 2010, most instances of insecurity that were reported concerned domestic disputes, although in March 2010 it was reported that a student was murdered in a dispute between rival cult gangs. Likewise, notable events in 2011 included the February reported killing of another student in a cult-related incident as well as a protest in April after the close of elections for governors and deputy governors. In August 2012, nearly 40 lawyers barricaded magistrates' courts to protest the abduction of a newly appointed judge. In December 2012, the mother of the Minister of Finance was reportedly kidnapped for ransom in Aniocha South. Violence around kidnappings and robberies increased in 2013, resulting in several reported deaths throughout the year. In 2013, according to Vanguard newspaper, there was some confusion over the Aniocha North council budget leading to tension among the committee members. There were also two reported incidents of bank robberies leading to the death of several suspected criminals.

2.5.4 Warri North/South/Southwest

Throughout the time period examined, most incidents of insecurity and violence in Warri concerned attacks on oil pipelines and infrastructure, kidnappings and abductions, robberies, and politically-motivated violence related to the Amnesty Law provisions and elections. In March 2009, suspected militants sabotaged an oil pipeline near the Warri River which reportedly led to a police shootout. Later that month, it was reported that at least one oil thief was killed when men in speedboats opened fired on a patrol of JTF officers in Warri South. In June, it was reported that four militants reportedly with MEND were killed after they attacked and destroyed the Makaraba-Utonana-Abiteye pipeline operated by Chevron. In August, a passenger boat, reportedly belonging to the Itsekiri community was attacked by armed Ijaw youth, allegedly killing one passenger while others were robbed. Following the attack, it was reported that tension between Itsekiri and Ijaw ethnic groups along the Escravos area escalated. In 2010, the year opened with reports of energy workers raiding a building owned by their employers in protest of treatment and wages. In March, it was reported that MEND detonated twin car bombs near a government building reportedly killing one person and injuring six others. It was reported that a dialogue concerning the Amnesty law and provisions was being held with four Nigerian governors present, although none were harmed. In April and May, it was reported that hundreds of youth protested charges brought against a former Delta state governor over corruption. In November, it was reported that youth unemployment and criminality was spiraling due to inadequacies in the Amnesty program jobs training program. In 2011, violence and incidents of insecurity concerning robberies and other criminal activities that plagued Warri in the previous year persisted. In the first half of the year, there were several reports of robberies that resulted in lynching and suspects being burned alive. In the second half of the year, it was reported in August that unidentified gunmen blew up Dibi Flow Station in Warri North while in December a prominent community leader was murdered by a youth gang in what appeared to be an attempted robbery. In a controversial move, Delta state banned motorcycle taxis in Warri, which raised concerns of protests over unemployment and limited transportation options. In 2013, gunmen reportedly killed a lecturer at the Petroleum Training Institute (PTI). In July, youths armed with rocket propelled grenades, submachine guns, and explosives reportedly invaded several communities and killed 12 people. Some feared this incident would lead to increased ethnic tension but representatives of the Ijaw and Itsekiri communities immediately acted to de-escalate the situation.

2.5.5 Udu

From 2009 to 2010, there were only three reported incidents in Udu, two concerning robbery and domestic disputes while hundreds of students from the Delta Steel Company Technical School reportedly protested an increase of school fees in

January 2010. In 2011, it was reported in March that police broke up a kidnapping gang and rescued two children belonging to a prominent leader of the Petroleum Tanker Driver's Association. In May, it was reported that the JTF launched a manhunt for the leader of the NDLF, reportedly resulting in his death later that month. The year closed in November with the report that a candidate for the governorship was killed by police after one of his aides was arrested for possession of weapons and ammunition. In early 2012, an alleged thief was reportedly lynched by vigilantes. Cult violence was also reported. In 2013, there were multiple reports of violence related to armed robbery and police clashes with criminals. There also continued to be reports of deaths and mutilations related to cult activity.

2.6 Cross River State

Incidents Per Capita Rank 8/37; Fatalities Per Capita Rank 19/37 (Fig. 2.7).

To the southeast of Nigeria, the coastal state of Cross River is home to approximately 2.9 million people (2006 census), predominantly of Efik, Ejagham and Bekwarra background. One of the fastest growing states in Nigeria, Cross River is endowed with vast mineral resources, plentiful arable land, and a growing number of tourist attractions. Liyel Imoke, of the People's Democratic Party (PDP), was elected governor of Cross River in August 2008 after his first electoral victory of April 2007 was annulled by an Election Appeal Tribunal. He was re-elected in February 2012 (www.crossriverstate.gov.ng).

For years, Cross River was the stage to a heated territorial dispute between Nigeria and Cameroon over the oil-rich Bakassi peninsula. After a controversial UN-backed ICJ verdict in 2002 and a comprehensive resolution between the two nations in 2006, Abuja began to transfer authority of the peninsula to Yaoundé, and Cameroon eventually took full sovereignty of Bakassi in August 2013.

Otherwise, after three relatively peaceful years from 2009 to 2011, Cross River saw an increase in violence in 2012–2013, with two notable peaks in the first half of 2012 and first half of 2013. Overall, 47 violent incidents were reported that led to the deaths of over 170 people, particularly around the capital city of Calabar to

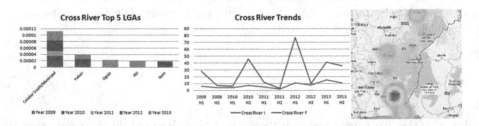

Fig. 2.7 Bar Chart shows annual per capita incidents for Top LGAs; Line Graph shows trend in # of incidents and fatalities in the state; Heat Map shows hotspots of violence from 2009 to 2013—Nigeria Watch data formatted and uploaded to Peace Map (www.p4p-nigerdelta.org/peace-building-map)

the south and in the Yakurr, Ogoja and Abi Local Government areas (LGAs). While the nature of violence in the capital varies, land competition and communal clashes remain the primary causes of fatalities in LGAs outside of Calabar according to the data. The following LGA-level breakdown describes risk factors in the five most violence-prone LGAs in the state. The incidents described are not exhaustive but give a brief overview of the scope and tenor of the types of issues reported in the data during the five year period of 2009–2013.

2.6.1 Calabar Municipal South

The largest and capital city of Cross River, Calabar has experienced the highest levels of per capita violence in the state in the period of 2012–2013, although there were some incidents of violence and general insecurity from 2009 to 2011. Predominant issues related to crime, domestic violence, piracy, as well as clashes between gangs, cults or political groups. In 2009, of the six reports of insecurity, five had to do with domestic disputes or attempted robberies. In December, however, there were reports that the Nigerian Navy clashed with suspected pirates that were alleged members of the Bakassi Freedom Fighters gang. It was reported that three of the suspects were killed. In 2010, significant events included the murder of a television crewman attached to the Government House press corps after a dispute in a bar in May. In June, union-sponsored campus protests took place at the University of Calabar over unpaid salaries which reportedly closed the university temporarily. In mid-October, a clash between Nsadop and Boje communities, over land rights, reportedly killed 35 people, including a soldier. By the end of the month, it was reported that a dispute between the communities had resulted in the displacement of thousands while causing a sharp decline in tourism, which is a mainstay of Calabar's economy. In 2011, the main incidents of insecurity reported concerned alleged murders and abductions by cult gangs. In June, two rival cult groups reportedly clashed over territory and control, resulting in three deaths. In March 2012, four were reportedly killed in a clash between Vikings and KKK members. In June 2012, six were reportedly killed in a similar clash. Additionally, in September 2012, four were reportedly killed in a PDP intra-party dispute. Furthermore, as the city is the political capital of the state, there were a number of protests in 2012–2013, led either by employees of the state's internal revenue service (September 2012), university students and staff (September 2012, August 2013, October 2013), or those protesting over the Bakassi issue (October 2012). Finally, the city experienced a landslide in September 2013, reportedly killing 10.

2.6.2 Yakurr

From 2009 to 2011, there were only three incidents reported in Yakurr, the most significant occurring in March of 2009 when it was reported that more than 5,000 armed men stormed the Nko community in an inter-communal land dispute, razing homes and killing a reported 15 residents. It was also reported that following this incident, many residents fled to Calabar and other neighboring cities. Apart from a robbery that led to the death of three in Ugep in January 2012, the LGA has seen a handful of continuing inter-communal land disputes. In April 2013, eight were reportedly killed in a clash between two communities over a piece of farmland. In June 2013, four people in one community were reportedly killed by the people of another after a suspect was apprehended for allegedly stealing.

2.6.3 Ogoja

From 2009 to 2010, there were three incidents reported in Ogoja, with an inter-communal land clash in March of 2010 reportedly leading to the death of several combatants. There were no reported incidents in 2011, although the LGA was hit by heavy rainstorms in May 2012, reportedly killing three and displacing thousands. Additionally, there were issues of crime and vigilante justice in 2013, a deadly clash between youths and police in October 2013, and a violent clash between two rival cult groups in November 2013.

2.6.4 Abi

From 2009 to 2012, there was only one reported incident involving the arrest of militia members in April 2010. From 2012 to 2013, however, long-standing land disagreements turned violent. In January 2013, a community in Ikwo LGA in neighboring Ebonyi State reportedly clashed with communities in Abi. During this incident over a dozen people were reportedly killed in the course of a week. A similar clash led to seven deaths in March 2013. Separately, a police inspector was also killed by armed robbers in September 2013.

2.6.5 Ikom

With only two reported incidents from 2009 to 2010, Abi remained relatively peaceful with the exception of a longstanding conflict between two border communities over farmland that reportedly claimed the lives of five persons in March 2009. Tensions flared again in October 2010, reportedly resulting in the displacement of hundreds of people. There were no reported incidents in 2011but in August 2012, members of a Bakassi Self Determination Front reportedly declared their independence from Nigeria. In May 2013, a traditional ruler and one other person were reportedly kidnapped by unknown gunmen who demanded 20 million Naira for their release.

2.7   Edo State

Incidents Per Capita Rank 8/37; Fatalities Per Capita Rank 11/32 (Fig. 2.8).

Landlocked between Ondo, Kogi and Delta States, Edo is home to about 3.2 million people (2006 census), predominantly of Edo, Bini, Owan, Esan, and Afemai background. Edo's economy centers on agriculture, including food crops such as yams, cassava, rice or maize and cash crops such as rubber, palm oil, cotton, cocoa and timber. The State's capital, Benin City, is the center of Nigeria's rubber industry. Edo also contains significant deposits of granite, limestone, marble, lignite, crude oil, gold, and kaolin clay.

Edo's State governor, Adams Aliyu Oshiomhole, assumed office in November 2008 after winning an appeal in the 2007 elections, which had initially declared his rival Oserheimen Osunbor governor. In July 2012, Oshiomhole was reelected for a second term in a landslide victory. He is one of six governors affiliated with the Action Congress of Nigeria Party (ACN) (www.edostate.gov.ng).

In the time period examined, Edo was Niger Delta's eighth most violent state on a per capita basis, with 184 incidents claiming the lives of 451 people. Issues ranged from protests, criminality, abductions and domestic violence to clashes between gangs, cults, political groups or communities. Over half of these incidents were reported in the Oredo Local Government Area (LGA), home to Benin City, although violence was also reported further north, notably in the Esan West, Uhunmwonde, and Etsako Central, East and West LGAs. The following LGA-level breakdown describes risk factors in the five most violence-prone LGAs in the state. The incidents described are not exhaustive but give an brief overview of the scope and tenor of the types of issues reported in the data during the five year period of 2009–2013.

2.7.1   Oredo

Home to Benin City, Oredo LGA experienced collective violence between gangs, cults, religious groups, members of political parties or communities. In 2009, several instances of cult-related violence were reported, in one case reportedly

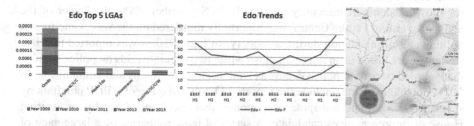

Fig. 2.8 Bar Chart shows annual per capita incidents for Top LGAs; Line Graph shows trend in # of incidents and fatalities in the state; Heat Map shows hotspots of violence from 2009 to 2013—Nigeria Watch data formatted and uploaded to Peace Map (www.p4p-nigerdelta.org/peace-building-map)

resulting in seven deaths in February. Also that year, a conflict between University of Benin and University of Lagos students reportedly left at least three dead and many others injured in November. In 2010, cult violence, robberies and domestic disputes led to over 50 deaths, including eight people in reported cultist clashes in March. Also during the year, it was reported in September that hundreds of youth took to the streets of Benin City to protest the failure of the state governor to reconstitute the board of the Edo State Oil and Gas Commission and, in December, two students were killed at a protest over the postponement of final examinations. 2011 was also characterized by much of the same types of violence, including a dispute in May between two rival cult groups that reportedly led to the death of six people. There were also reports of kidnappings and abductions throughout the year, although the number of fatalities could not be confirmed. In January 2012, an inter-cultist clash between Eiye and Black Axe Confraternities reportedly killed eight over the course of a week, while two similar clashes in June and July 2012 led to the deaths of three and four respectively. The same Eiye Confraternity also lost some of its members through an intra-cult clash in November 2013. Ethnic/religious groups also clashed in 2012–2013, notably in early 2012 when two mosques and an Islamic school were attacked, killing five and forcing many to flee. A pastor was also killed in November 2013. Separately, in May 2012 the Principal Private Secretary to Edo State's Governor was murdered at his residence. In April 2013, factions of the PDP and the ACN clashed during the local government elections. In July 2013 the Deputy National Chairman of the All Nigeria Peoples Party (ANPP) was attacked at his home. There were several cases of abductions reported in Oroedo, including that of an Israeli expatriate in July 2013 and of three female teachers in August 2013. Finally, as the political capital of the state, there were multiple protests in Oredo in 2012–2013. In January 2012 there was a protest against a fuel subsidy removal. In August 2012, there was a protest for the release of a human rights activist. In March 2013 there was a protest against levies and extortion from the Road Transport Employees Association of Nigeria (RTEAN). In June 2013 there was a protest after the alleged killing of a student by police.

2.7.2   Etsako Central/East/West

From 2009 to 2011, armed robberies, cult violence, and domestic disputes accounted for most incidents of insecurity and death. In September 2009, the leader of the legislative arm of Etsako Central was reportedly murdered on the way home from a holiday at a filling station during a robbery. In late 2010, it was reported that five students were killed in an inter-cult clash while two separate raids on police stations that year led to several police deaths as well as stolen weapons and ammunition. In 2011, notable incidents include the reported October slaying of a PDP chieftain by suspected cultists and a dispute over money that was handed out by an aspiring House of Representatives candidate. A series of bank robberies by a large gang of armed gunmen reportedly led to the deaths of over a dozen people in Etsako West in November 2012. In April 2013, around the time of the local government elections, a clash between supporters of two parties reportedly led to two fatalities. PDP supporters reportedly protested the results of the election, calling for another vote. The All Progressive Congress (APC) chief in Edo was also kidnapped in August 2013, but eventually released in September 2013 and his kidnappers arrested.

2.7.3   Akoko Edo

From 2009 to 2011, robberies and attempted robberies accounted for most of the fatalities and injuries in Akoko Edo. In November 2009, three policemen were shot dead during a bank robbery while in June 2010, two more police fatalities were reported after an attack on a police station. In January 2011, primary election-related violence reportedly killed one while robbers released by police in November were killed in an attack by an incensed mob. From 2012 to 2013, several deaths reportedly occurred during domestic disputes while ritualistic killing and robberies also accounted for at least two deaths and multiple injuries.

2.7.4   Orhionmwon

With less than 200,000 people according to the 2006 census, Orhionmwon reported only nine incidents from 2009 to 2013, with the most deaths occurring in February 2009 when an ambush on a Central Bank bullion truck reportedly led to seven deaths, including one of the robbers and one police officer. In 2010, it was reported that the chief of Igbekhue kingdom was murdered after a robbery attempt on his warehouse while over 100 youths gathered in June to protest the failure of the Niger Delta Development Commission to fulfill a roads project contract. In 2011, election-related violence reportedly led to at least six deaths, including a PDP youth leader in April. Also during the year, there was a boundary dispute between the Dumu-Igbo communities of Delta state and Evbohighae community of Edo State, although there were no reports of violence or deaths. In August 2013, it was reported that six people were kidnapped for ransom in one incident while a protest occurred in October over the abandonment of a road construction project. There were no reported fatalities associated with either incident.

2.7.5   Esan Northeast/Southwest/Central/West

In late 2009, in an operation named Attack to Rescue, it was reported that police killed six robbers while successfully rescuing a kidnap victim. In 2010, a February riot occurred at Ambrose Alli University over proposed tuition hikes while at least seven students were reportedly killed at the same university in October in a rival gang shoot-out. In 2011, domestic clashes and flooding accounted for at least four deaths, according to reports. Esan West was relatively violent in 2012–2013. There were two violent cult clashes in early and late August 2012, leaving two and three dead, respectively. The killing of an ACN official was reported in April 2013, as well as that of a prominent businessman in October 2013.

2.8 Imo State

Incidents Per Capita Rank 20/37; Fatalities Per Capita Rank 28/37 (Fig. 2.9).

Imo state has a population of approximately 3.9 million people, according to the 2006 census. The population of Imo state is predominantly Igbo (98 %). The capital city of Owerri is the largest in the state. Imo is made up of twenty-seven Local Government Areas (LGAs).

Imo's economy mainly consists of exporting natural resources such as palm oil, mahogany, crude oil, and natural gas. Due to the high population density and over-farming, the quality of the soil is reportedly worsening.

Owelle Rochas Okorocha has been the governor of Imo since May 2011. In 2011, he left the People's Democratic Party (PDP) to run for governor with the All Progressives Grand Alliance (APGA). The Independent National Electoral Commission (INEC) initially declared the election inconclusive due to reports of irregularities. After being elected, in a controversial move, Governor Okorocha fired the local government chairmen and replaced them with a transition committee (en.wikipedia.org/wiki/Imo_State).

Violence per capita in Imo is among the lowest in the region, as is the number of fatalities per capita. Incidences of violence largely occurred in the LGAs surrounding the capital city of Owerri. Between January 2009 and December 2013, incidents reported included criminality, abductions and vigilante/mob justice. There were also a number of fatalities associated with public unrest and reports of ritual killings in the state. In 2011, insecurity, mainly related to kidnappings and attempted kidnappings, spiked in Owerri and Ohaji/Egbema, although it decreased in subsequent years. The following LGA-level breakdown describes risk factors in the five most violence-prone LGAs in the state. The incidents described are not exhaustive but give a brief overview of the scope and tenor of the types of issues reported in the data during the five year period of 2009–2013.

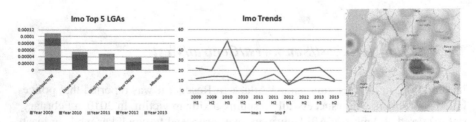

Fig. 2.9 Bar Chart shows annual per capita incidents for Top LGAs; Line Graph shows trend in # of incidents and fatalities in the state; Heat Map shows hotspots of violence from 2009 to 2013— Nigeria Watch data formatted and uploaded to Peace Map (www.p4p-nigerdelta.org/peace-building-map)

2.8.1 Owerri Municipal North/West

In 2009 and 2010, violence in Owerri mainly concerned domestic disputes and attempted robberies although kidnappings also became increasingly prevalent, starting in 2010. In March, it was reported that police had successfully rescued the State Ministry Deputy Director of Lands, who had been abducted earlier in the month, while also killing his kidnappers. Similarly, in April, four people were kidnapped who were visiting officials at the Nigeria Agency for Food and Drug Administration and Control (NAFCAC) and held for a ransom of four million Naira. Also, in January of 2010, a large street protest numbering in the hundreds reportedly marched on the Imo State Government House demanding the removal of then-President Yar'Adua. Insecurity during 2011 was also characterized by attempted kidnappings, with an attempted abduction of the brother of the Minister of the Interior in January followed by a reported kidnapping of a judge of the Nigerian High Court in February. In July, violent demonstrations were reported at Federal Polytechnic Nekede over an increase in school fees which reportedly left students dead. Also in 2011, three members of the Movement for the Actualization of the Sovereign State of Biafra (MASSOB) were reportedly killed by police and 60 others were injured at a celebration of 44 years since the declaration of Biafran independence. At the same event, it was also reported that up to 300 MASSOB activists were arrested and detained by police.

Between January 2012 and December 2013, the LGAs around Owerri had the highest number of reported incidents of insecurity per capita in the state. There were a few lynchings and attempted lynchings of suspected robbers reported, several kidnappings, including the abduction of a popular actress, and a number of other murders, including a couple of ritual killings.

There were both peaceful protests and violent riots reported as well. In April 2012 it was reported that indigenes protested a government land seizure intended for development. There was also a clash of rival cult groups reportedly leaving several dead in December 2012. In 2013, there was political tension as the local government chairmen that Governor Okorocha had fired in 2011 protested peacefully in the streets, complaining that the democratic process was being undermined. Also in 2013 there were more reported clashes between police and individuals and gangs suspected to be kidnappers. There was an increase in the number of student protests in the early months of 2013. First, in January, a reported 2,000 youths took to the streets to protest violence in Owerri. In February another protest was staged by youths apparently angered over aspects of an amnesty agreement reached with former President Yar'Adua. Also, in October, thousands of women reportedly took part in a peaceful demonstration protesting alleged intimidation and harassment of state government officials by federal anti-corruption agencies. Specifically, according to the Vanguard newspaper, the women alleged that the Economic and Financial Crimes Commission (EFCC) and the Independent Corrupt Practices Commission (ICPC) had been unfairly targeting the administration of Gov Rochas Okorocha for political reasons.

2.8.2 Ehime-Mbano

Criminality and violent interpersonal conflict were the main issues in Ehime-Mbano during this period, including a bank robbery and several reported murders. In 2009 and 2010, there were four reported murders, including a Port Harcourt-based businessmen kidnapped for ransom, a former local political leader and an aspiring House Assembly speaker. Also, in November 2010, the monarch of Umunakanu Ezeala community was reportedly murdered on his way to a meeting with a popular local indigene. It was reported that the Police State Command took responsibility for the killing although it was unclear the motive.

In late 2013, there was a reported incident of intra-communal conflict that led to the death of at least two people, suspected to be related to a land dispute. In November, it was reported that youths took to the street to protest unfulfilled political promises in a rally that turned violent, with the death of a police officer as well as property damage reported. Issues such as the high rate of youth unemployment and political marginalization were cited in an interview with one self-described "youth leader" who participated in the protest.

2.8.3 Ohaji/Egbema

Ohaji/Egbema, with a population of approximately 182,000 according to the 2006 census, reported two to three incidents of insecurity per year from 2009 to 2013. Several incidents revolved around clashes between herdsmen and farmers who protested the free grazing of cattle on their land. In early 2009, an oil spill resulted in an explosion which reportedly killed ten people who were trying to gather the spilled oil into tankers. Then, in January 2010, a reported clash between two ethnic groups over the killing of a woman left houses destroyed and people injured. In March, a clash over the election results of a youth association, including the sharing and distribution of oil products, reportedly left at least three people dead and property damaged. In July 2011, it was reported that Fulani herdsmen entered a particular village to graze cattle and destroyed crops, leading to clashes that killed at least two and injured others. Also that year, cult-related violence, foiled kidnapping attempts, and armed robbery reportedly also accounted for at least a half dozen deaths. In October 2012, flooding caused a spike in food prices, leading to reports of displacement while in April 2013, up to 1,000 people reportedly gathered to protest attacks and invasions by pastoralists on village farmlands.

2.8.4 Ngor-Okpala

Conflict risk factors during this time period in Ngor-Okpala were mainly related to criminality, including kidnapping, murder, and ritual killing. In 2009, a local politician was reported murdered while in 2012 and 2013, a ritual killing and the killing of a notorious kidnapper were the main incidents reported.

2.8.5 Mbaitoli

From 2009 to 2011, there were five reported incidents of insecurity in Mbaitoli, an LGA of approximately 240,000 people. Three incidents were speculated to be tied to kidnapping and ransom, although there were no further reports confirming or denying these allegations. The other incidents reported concerned a contract dispute and a stabbing death, with no motive reported.

In April 2012, a police station was attacked by gunmen. In a separate incident in May, a monarch was reportedly killed by gunmen. In January 2013, the deputy governor's director of protocol was reportedly murdered by gunmen who later claimed the attack was not politically motivated. In August, the alleged leader of an armed robbery gang was reportedly killed by the police during an exchange of gunfire while others were arrested in connection with a variety of robberies and other criminal activities.

2.8.6 Other LGAs

2.8.6.1 Oguta

Oguta is on the east bank of Oguta Lake of approximately 142,000, according to the 2006 census. The city, and the LGA named after it, thrives on tourism as well as commercial fishing. The town is not usually prone to violence, although there was an increase in the first half of 2013.

From 2009 to 2011, there was only one report of insecurity, a dispute over an *ezeship* (traditional leadership) that led to a murder that was purportedly contracted through an assassin, although there were no further details confirming this report. October 2012 was marked by floods that displaced more than 8,000 people, destroyed crops and livelihoods, and reportedly caused an increase in food prices in the months that followed. In early 2013, there were reports of abductions and killings of hotels' managers and businessmen. In a sign of increased political tension in Imo, thugs reportedly attacked the governor's convoy in June 2013. There were also several reported deaths towards the end of the year related to foiled robbery attempts and clashes between police and suspected robbers and kidnappers.

2.9 Ondo State

Incidents Per Capita Rank 28/37; Fatalities Per Capita Rank 32/37 (Fig. 2.10).

Ondo state has a population of approximately 3.44 million according to the most recent census (2006). The majority are of Yoruba descent, with a sizable minority of those from Ijaw subgroups, particularly along the coast. Ondo derives most of its revenue from the production of cocoa, palm oil, rubber, lumber, and cassava. Approximately 65 % of the labor force is employed in the agrarian sector. The state is also rich in oil and minerals (www.ondostate.gov.ng/new).

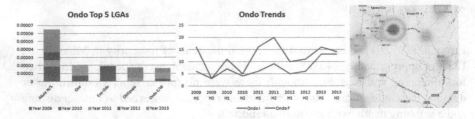

Fig. 2.10 Bar Chart shows annual per capita incidents for Top LGAs; Line Graph shows trend in # of incidents and fatalities in the state; Heat Map shows hotspots of violence from 2009 to 2013—Nigeria Watch data formatted and uploaded to Peace Map (www.p4p-nigerdelta.org/peace-building-map)

On a per capita basis, violence in Ondo was relatively low in comparison to the other Niger Delta states according to Nigeria Watch data. It did, however, see a gradual increase in reported insecurity throughout 2012–2013.

Governor Olusegan Mimiko, Labour Party (LP), who was re-elected in October 2012, subsequently announced plans to build refineries, petrochemical and fertilizer plants and signed a $3.3 billion deal with Dangote Group, the largest manufacturing conglomerate in West Africa. After the 2012 gubernatorial election, the losing party raised concerns about alleged election irregularities and intimidation.

Overall, much of the violence in Ondo state was related to instances of petty crime and some gang violence, although fatalities associated with kidnappings for ransom appeared to be on the rise. The most reported incidents of violence occurred in Akure N/S, the largest city and capital of Ondo state. The following LGA-level breakdown describes risk factors in the five most violence-prone LGAs in the state. The incidents described are not exhaustive but give a brief overview of the scope and tenor of the types of issues reported in the data during the five year period of 2009–2013.

2.9.1 Akure North/South

Akure is the capital and largest city in Ondo state. It is the main trading center in the state for agricultural produce, including cassava, yams, corn, cotton and tobacco. While there were no incidents reported in 2009, in February of 2010, five people were reported killed in a series of bank robberies across the city. In 2011, a reported chieftaincy rivalry led to the arson of the home of a deposed ruler. While an exact number of deaths were not reported, allegedly the wife and at least once child perished in the house fire. Incidents of reported insecurity throughout 2012–2013 increased during this period. During the second half of 2012, violence sometimes had a political dimension, in the context of a hotly contested gubernatorial election in October 2012. Ondo is the only state in Nigeria controlled by the Labour Party. Both the ACN and the PDP filed appeals contesting the outcome, however the Supreme Court ruled that the election was valid. Gang violence in Akure increased around the time of the election and continued throughout 2013. The Ade Basket Boys is reportedly among the most active gangs in Akure and is believed to have been involved in riots leading up to the election as well as various criminal activities, including armed robbery.

Between March and May of 2012, eight prominent people were reportedly kidnapped for ransom in Akure, including family members of politicians and government officials, a journalist, and a businessman.

In the summer of 2013, there was a prison break in Akure when gang members freed an estimated 175 prisoners and fatally shot two civilians. That same week, Nigerian Immigration Services arrested and deported 147 allegedly illegal immigrants from Niger and Chad. State Comptroller Mr. Sola Sessi stated that the presence of those immigrants in Ondo presented a threat to state security and said similar operations will continue. In the second half of 2013, students at the Federal University of Technology Akure staged at least two protests. One protest was in response to the alleged rape of three female students. The other demonstration was in protest of a teacher strike.

Other reported issues in 2013 included several murders, a peaceful protest by nurses over an alleged assault, and a case of police corruption.

2.9.2   Ose

While there were no reported incidents in Ose in 2009, violence increased from 2010 to 2011, with a reported seven fatalities during this time period. In 2010, an armed robbery allegedly led to the death of two constables, with the suspected arrested by police a short time later. In March 2011, a clash between supporters of the ACN and LP led to the death of one person, reportedly killed by a stray bullet. In October of the same year, a robbery attempt led to the alleged death of at least three people, two suspects and one police officer. There were no incidents reported in 2012 to 2013.

2.9.3   Ese-Odo

Over the time period examined, there were three reported incidents in Ese-Odo, allegedly leading to the deaths of 11 people. In February 2009, a conflict of a disputed chieftaincy erupted between two communities, leading to the deaths of at least six people and the destruction of traditional structures, such as the monarch's palace and several surrounding buildings. Later, in April, it was reported that four fisherman had been killed during a protest by private security forces working for a Nigerian oil exploration firm. There were no reported incidents from 2010 to 2011. In 2012, however, the death of a suspected kidnapper in police custody was reported after his family allegedly began petitioning the governor to look into the circumstances surrounding their son's death. There were no reported incidents in 2013.

2.9.4   Okitipupa

There were three incidents reported in Okitipupa during the time period examined. The first, in October 2011, a motorbike driver was reportedly shot by a police officer after refusing to pay a bribe. The following day, other motorbike drivers marched on the police station, protesting the killing. One person was allegedly killed during the protest. Violence spiked again in 2013 with an incident of domestic violence in January resulting in the alleged murder of one woman. In March, a botched bank robbery reportedly led to the death of at least one police officer and one suspect. Also, in August, a young man was found hanged under suspicious circumstances, although no formal charges appeared to have been brought against any suspects by the year's end.

2.9.5   Ondo East/West

While an attack on the National Assembly building in June 2010 led to the reported deaths of two security guards, insecurity spiked in 2011, mostly around domestic incidents, robberies and cult-related violence. In December 2011, a clash between rival cult groups reportedly led to the death of at least five people, with several others injured. There were no reported incidents in 2012, although violence spiked again in 2013 with the reported killing of a college student after an altercation with a police officer in February. In August of the same year, the teenage son of prominent union leader was allegedly murdered outside of the family residence, although no immediate cause or suspect was reported at the time.

Chapter 3
North Central Overview

While grouping states regionally can often help disaggregate the various conflict ecosystems in the country for more analytical clarity, in this case the variations are perhaps more disparate than in other regions. The two North Central states of Jigawa and Katsina were among the most peaceful in the country during the period of 2009–2013. Kano and Kaduna, however, had significant spikes of violence during the period. Kaduna itself has a high degree of complexity and variation, with inter-communal violence in the southern part of the state, as well as high levels of election violence in 2011 and terrorism in 2012. With the loss of candidate Muhammadu Buhari, a northerner from the Congress for Progressive Change, to incumbent Goodluck Jonathan, a southerner from the Niger Delta, during the presidential elections, violent riots broke out and led to sectarian killings, with Muslim rioters killing Christians and members of ethnic groups from southern Nigeria, and Christians retaliating by killing Muslims and burning mosques and other properties. In the predominantly Christian towns of southern Kaduna, for example, violence left more than 500 dead, the majority being Muslim. Despite the police managing to protect Muslims and Christians who had fled to police stations for safety, they could not control the surge of violence occurring outside the stations and barracks. Unfortunately, the eventual subduing of protestors and mobs ended in reported human rights violations due to excessive use of force by the police (Nigeria: Post-Election Violence Killed 800 2011).

Although the command and control of the so-called "Boko Haram" phenomenon is frequently ambiguous, in 2012 JAS did claim responsibility for some attacks outside of their base in the Northeast, including bombings in Kaduna such as a suicide bombing of churches on Easter, the suicide bombings of the *This Day* newspaper offices on April 26, the bombings of multiple churches on June 17, (which then instigated a backlash of violence between Muslims and Christians for a week), the attempted assassination of traditional Muslim leaders in July, and more church killings in July and August.

Kaduna experienced moderately high levels of armed violence in the years analyzed, with particularly violent outbreaks occurring between May 2011 and August

© Springer International Publishing Switzerland 2015

P. Taft, N. Haken, *Violence in Nigeria*, Terrorism, Security,
and Computation, DOI 10.1007/978-3-319-14935-6_3

2013. Armed violence in Kaduna seems at times to have had more to do with local politics than a wider ideological jihadist agenda. In particular, issues surrounding the rights of "indigenes" versus so-called "settlers" has led to communal violence in the state, similar to patterns witnessed in the Middle Belt. In addition, the levels of political violence in Kaduna have been historically higher than in other states, having the unfortunate distinction of having had the most destructive cycles of violence (in terms of lives lost as well as property damaged) since the end of the civil war in the 1960s (Action on Armed Violence 2014). As noted above, this violence often takes on an ethno-religious element, but is exacerbated by pressures from population growth as well as fluctuations in the world economy that lead to dips and peaks in the agricultural sector, which provides the main form of employment in the state. Organized violence, again often grouped along ethno-religious lines, and participation in criminal gangs or communal militias is often seen as a lucrative alternative for a population with a high percentage of unemployed young men. The recent influx of groups and individuals with ties to religious extremist groups, or sympathetic with their agendas, also provide a fertile ground for recruitment of armed thugs, paid to perpetrate acts of violence. According to some studies on violence in the state, the single largest factor fueling instability and violent attacks relates back to issues of governance, which affects everything from land issues to public faith in the security sector (IBID). Although typically governance tensions stem from the larger, historical, fractionalization between Christians and Muslims in the state, higher levels of insecurity due to the presence and increased attacks of gangs assumed to be affiliated with Boko Haram has also led to concerns about the ability of the government to provide basic protection to citizens. The rise of citizen vigilante groups, especially in rural areas, to combat criminal gangs and those thought to be associated with religious extremist groups may also be indicative of government challenges as well as the level of community trust for state security agencies, especially the police (Musa 2012).

Meanwhile, the capital of Kano State, the second largest city in the country, has been a major target of insurgent/terrorist attacks, beginning with the original uprising in 2009 which reportedly left 700 dead between security forces, militants and innocent bystanders (United States Department of State Publication 2010). Violence spiked again in 2012 starting with attacks on January 20, in which a group of gunmen entered multiple police buildings to free inmates followed by the bombing of eight government sites. Insurgents continued to fire upon police officers and random pedestrians in protest of the detention of Boko Haram members. This single coordinated attack reportedly left at least 150 dead, with the government implementing a 24-hour curfew (Elbagir and John 2012) and ordering hospitals to provide victims with maximum medical attention free of charge (Xinhua 2012). In March 2013, worldwide media attention focused on Kano and Boko Haram following a multiple suicide bomb attack on a bus station, which left a reported 20 dead and dozens more injured. Then in July, a car bombing in the Kano city neighborhood of Sabon Gari, a target thought to be chosen because of its high population of Christians as well as ex-patriate bars and nightclubs, also captured news headlines, particularly as it was reported that attacks singling out Christians and foreigners seemed to be on the rise (Hilton and Ndukong 2013).

In addition to insurgent and terrorist attacks, Kano has also been vulnerable to violence related to political and economic issues. Like Kaduna, Kano has experienced fluctuations in unemployment rates as well as widespread income inequality that has been associated with a rise in youth violence and participation in criminal activities (Human Rights Watch 2005). In January 2012, as in other parts of the country, the removal of the fuel subsidy led to demonstrations and protests that turned violent, with a BBC report citing approximately 300 injured in clashes between police and protestors (BBC news Africa 2012).

Despite these incidents, Kano State has historically been quite peaceful and much less volatile than neighboring states, including Kaduna. Nevertheless, in 2012, a new radical jihadist insurgent group emerged in and around Kano called Jama'atu Ansarul Muslimina fi Biladis Sudan (JAMBS or Ansaru). According to some reports, this splinter group diverged in reaction to Boko Haram's more indiscriminate violence in an attempt to realign the insurgency's strategic approach and ideological vision, as well as due to the rejection of Shekau's leadership. The new leader of the Ansaru faction, Abu Usmatul al-Ansar released a video in which he states that "Islam forbids [the] killing of innocent people including non-Muslims. This is our belief and we stand for it." (http://africajournalismtheworld.com/) Ansaru has reportedly been responsible for the kidnapping of a French national in Katsina and several other foreigners in Bauchi who were later killed in March 2013.

Notwithstanding all the variations at the regional, state, and local levels, taking the region as a whole, one insight does suggest itself. If, as has been posited, underdevelopment and economic deprivation are key drivers of insurgency, poverty does not seem to be predictive of the location of attacks. According to the National Bureau of Statistics, Jigawa and Katsina are among the poorest states in the country, and yet they are also among the least violent. Kano and Kaduna, by contrast, are less poor and more violent. And even within those states, violence has tended to occur in the urban centers where incidents of poverty are lowest (Fig. 3.1).

3.1 Kaduna State

Incidents per capita rank 16/37; fatalities per capita rank 6/37 (see Fig. 3.2).

Kaduna state has a population of approximately 7.47 million people. About a third live in the cities of Kaduna, the state capital, or Zaria. Predominant ethnic groups are the Gbari, Hausa, Fulani, Kamuku, Kadara, Kurama and Bajjuu, but there are 36 other ethnic groups across the state. Hausa/Fulani Muslim communities primarily live in the northern part of the state, while Christians from various ethnic tribes tend to live in the cities and southern LGAs. Kaduna produces several crops for export, primarily cotton, peanuts, shea nuts and tobacco. Kaduna city is the country's largest textile manufacturing centre and has an oil refinery. Zaria, the second largest city, is a major industrial centre that produces textiles, cigarettes, paper products, and bicycles. Both cities host several universities and colleges, including the Nigerian Defense Academy, as well as several agricultural and technical institutes (www.kadunastate.gov.ng).

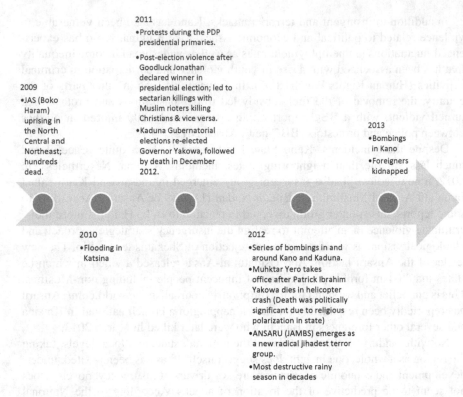

2011
- Protests during the PDP presidential primaries.
- Post-election violence after Goodluck Jonathan declared winner in presidential election; led to sectarian killings with Muslim rioters killing Christians & vice versa.
- Kaduna Gubernatorial elections re-elected Governor Yakowa, followed by death in December 2012.

2009
- JAS (Boko Haram) uprising in the North Central and Northeast; hundreds dead.

2013
- Bombings in Kano
- Foreigners kidnapped

2010
- Flooding in Katsina

2012
- Series of bombings in and around Kano and Kaduna.
- Muhktar Yero takes office after Patrick Ibrahim Yakowa dies in helicopter crash (Death was politically significant due to religious polarization in state)
- ANSARU (JAMBS) emerges as a new radical jihadest terror group.
- Most destructive rainy season in decades

Fig. 3.1   Timeline North Central, Nigeria

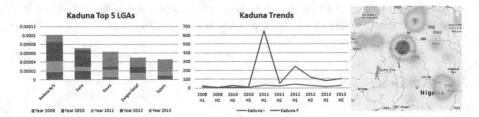

Fig. 3.2   Bar Chart shows annual per capita incidents for top LGAs; Line Graph shows trend in # of incidents and fatalities in the state; Heat Map shows hotspots of violence from 2009 to 2013—Nigeria Watch data formatted and uploaded to Peace Map (www.p4p-nigerdelta.org/peace-building-map)

The governorship of Kaduna has been in the hands of the PDP since the beginning of the Fourth Republic in 1999. The current governor of Kaduna, Muhktar Yero, took office on December 16, 2012, when then-governor Patrick Ibrahim Yakowa was killed in a helicopter crash while returning from a state funeral in Bayelsa. His death was politically significant because of religious polarization in the state, Yakowa being Christian and Yero, Muslim.

Although it reported the 16th highest rate of incidents per capita in the last four years, it reported the sixth highest fatality rate, indicating that incidents in Kaduna tended to be quite lethal. Kaduna experienced higher rates of violence in early 2011 and again in 2012. Although both years reported approximately the same number of incidents, fatalities were much higher during the election year of 2011. Violence decreased sharply in late 2012 and continued to drop into 2013, but is has yet to fall to the baseline level of 2009. Much of the violence during this period occurred in the state capital city of Kaduna. Political violence in response to presidential election results—which triggered a surge of extremist Islamist attacks—accounted for the increase in incidents and major spike in fatalities in Kaduna in early 2011. Violent rioting erupted around the country in 2011 when CPC and northern Muslim candidate, Muhammadu Buhari, lost the election. Violence was the most severe in the northern states; Kaduna in particular saw some of the heaviest casualties. Buhari supporters attacked people and burned Christian churches, homes, businesses and political offices. In retaliation, Christian supporters of Goodluck Jonathan burned mosques and killed Muslims. About 200 people were killed in the cities of Kaduna and Zaria, but many attacks took place in smaller towns and rural communities. Over 600 people were killed across the state. Kaduna also held gubernatorial elections in late April 2011, re-electing Governor Yakowa, before his death in December 2012. The following LGA-level breakdown describes risk factors in the five most violence-prone LGAs in the state. The incidents described are not exhaustive but give a brief overview of the scope and tenor of the types of issues reported in the data during the five year period of 2009–2013.

3.1.1 Kaduna North/South

Kaduna, the state capital, covers both Kaduna North and Kaduna South LGAs. Together, the city accounted for the highest number of fatalities and incidents, Kaduna North in particular. About half of the violence reported in Kaduna was attributed to the Boko Haram insurgency, targeting Christian churches, police and military installations, schools, and bars or nightclubs. Levels of violence were low in 2009 and 2010. In 2011, sectarian and political violence spiked dramatically, especially after the elections in April, when Goodluck Jonathan was declared the winner. Attacks intensified in 2012. On Easter weekend in 2012, a bomb blast killed over 100 people. In April 2012, insurgents also targeted several newspaper offices. In June 2012, over 70 people were reported dead and more than 300 were injured by attacks on churches, and then from subsequent retaliatory attacks that targeted mosques in Kaduna and Zaria. In late 2012, JTF agents uncovered and destroyed an arms cache believed to belong to Boko Haram. Five insurgents were killed in the process and the agents recovered bomb-making materials and weapons. There was also at least one incident of inter-communal violence reported that killed several in 2012. The majority of incidents reported during 2013 were shootings and armed clashes between insurgents, other gang members, and policemen.

3.1.2   Zaria

As in Kaduna North and South, violence was quite low in 2009 and 2010, although there were several reports of clashes between Sunni and Shia Muslims in 2009. In April 2011, post-election violence broke out when Goodluck Jonathan was declared the winner. This was followed by several incidents involving suspected insurgents. 2012 was by far the deadliest year for Zaria, more violent that the other three years combined. Many of the reported incidents in 2012 pertained to terrorist bombings and other attacks, the majority of which targeted churches. Suspected JAS also targeted bars and tried to bomb a motel frequently used as a brothel in October 2012. During the second half of 2012, police and JTF agents in Zaria engaged directly with suspected Islamists more frequently, resulting in fatalities of both insurgents and security forces. Police were also able to prevent a planned attack on a church in September 2012 and to recover weapons caches in November 2012. After Governor Yakowa died and was replaced by Yero, violence reduced dramatically.

3.1.3   Kaura

Bordering the Middle Belt state of Plateau, violence in Kaura LGA was primarily inter-communal in nature, unlike the more political and insurgent violence in Kaduna and Zaria LGAs. While Kaduna and Zaria experienced significant levels of violence around the election period and in the run-up to Governor Yakowa's death, Kaura by contrast remained relatively peaceful until 2013, which was by far the deadliest year for Kaura with almost twice as many fatalities in one year than in the preceding four. Although the state did not report as many total incidents, due to its small population the fatality per capita rate was almost equal to that in Zaria. Ethnic or religiously incited violence accounted for many of the incidents reported in Kaura. Many reports alleged that Fulani herdsmen attacked farming communities. In late March 2013, an attack on two villages left 22 dead and displaced an estimated 4,000 people.

3.1.4   Zangon Kataf

Very little violence was reported in Zangon Kataf LGA in 2009–2010. In 2011, some sectarian and election violence was reported. From 2011 to 2013, there were a number of reports of communal violence where suspected herdsmen attacked communities, often at night or in the early morning. These raids often targeted churches or local leaders. In March of 2013, mourners were attending a funeral for a traditional ruler when gunmen reportedly attacked and killed six.

3.1.5 Kajuru

As in several other Kaduna LGAs, election violence was reported in Kajuru in 2011. In 2012, a bomb blast was reported on a road near a mosque. In 2013, several incidents were reported including an attack by gunman on a police station that killed three.

3.2 Kano State

Incidents per capita rank 30/37; fatalities per capita rank 21/37 (see Fig. 3.3).

Located in northern Nigeria, Kano is the twentieth largest and most populous state in Nigeria. Its 9.4 million people (2006 census) are predominantly of Hausa and Fulani background. Approximately 75 % of the state's working population is engaged directly or indirectly in agriculture, with millet, cowpeas, sorghum, maize, and rice are the principal food crops for local consumption while groundnuts and cotton as the primary products for export and industrial purposes. Kano is also the largest commercial and industrial center in northern Nigeria, with numerous important markets and over 400 privately owned medium and small scale industries producing a plethora of products, from tanned leather to pharmaceuticals (www.ngex.com/nigeria/places/states/kano.htm).

Kano was Nigeria's twenty-first most violent state over the period January 2010–December 2013, with 162 incidents killing nearly 700 people. The state saw two spikes in violence in this period, the first and considerably larger being in the first half of 2012 and the second being in the first half of 2013. The first spike was a surge in insurgent activity and responses by the JTF, opened with a bombing attack by suspected Boko Haram in the state capital of Kano which killed 200+ people in January 2012. Conflict between insurgents and the JTF continued through the second half of 2012, albeit at a subdued level, before flaring up again in the first half of 2013, marked by a suspected Boko Haram bombing of an interstate bus terminal in March which killed around 70 people.

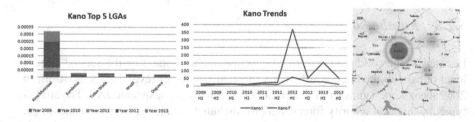

Fig. 3.3 Bar Chart shows annual per capita incidents for top LGAs; Line Graph shows trend in # of incidents and fatalities in the state; Heat Map shows hotspots of violence from 2009 to 2013—Nigeria Watch data formatted and uploaded to Peace Map (www.p4p-nigerdelta.org/peace-building-map)

In 1999, Rabiu Kwankwaso was elected as the first civilian governor of the Fourth Republic on the PDP Platform. He was replaced in 2003 by Governor Ibrahim Shekarau of the ANPP. In 2011, Kwankwaso was reelected as governor. The following LGA-level breakdown describes risk factors in the five most violence-prone LGAs in the state. The incidents described are not exhaustive but give a brief overview of the scope and tenor of the types of issues reported in the data during the five year period of 2009–2013.

3.2.1   Kano

Kano city, the capital of Kano state, stretches over a number of local government associations, including Dala, Fagge, Tarauni, Gwala, Ungongo, Kumbotso, and Nasarawa, as well as Kano's municipal LGA proper. In 2009 and 2010, violence mainly pertained to incidents of criminality and a few lynchings of suspected criminals. In 2011, post-election violence reportedly led to sectarian violence. Then, in 2012, violence spiked dramatically in the state with the city of Kano as the epicenter, experiencing the vast majority of incidents and an even greater proportion of the fatalities. In the first half of 2012 there was a deadly insurgent campaign, including an attack in January involving 20 bombs exploding throughout the city, which reportedly killed between 200 and 250 people, the vast majority of them civilians. Ten additional, unexploded bombs were found by security forces in the following few days. That bombing was the first of numerous incidents involving suspected JAS, although no subsequent attack was nearly as lethal. After the end of May, the violence declined significantly, and remained at a substantially lower level through the rest of the year. Violence picked up again in the first half of 2013, the largest incident of which was the suicide bombing of an interstate bus terminal that killed 70 people. The perpetrators were alleged to be members of JAS, though no one claimed responsibility. Another incident occurred at the end of July, when three coordinated explosions ripped through three different parts of the city, killing around 30 people. After this incident, however, violence tailed off, with few incidents the rest of the year.

3.2.2   Kumbotso

Located on the southern outskirts of the capital, Kumbotso saw the second highest level of per capita violence, albeit at a substantially lower level than Kano itself. In 2012 there were a number of violent incidents involving suspected JAS. The attacks typically targeted security forces, but also included the killing of two businessmen, and the bombing of a government-run Islamiyya primary school, the first such incident affecting an Islamic institution. Since the first half of 2012, however, the level of violence in the LGA significantly reduced, though there have been a number of attacks by unknown gunmen targeting security forces, schools, and a politician.

3.2.3 Tudun Wada

From 2009 to 2011 reported incidents mainly pertained to criminal and interpersonal violence. In March 2012, suspected JAS members set fire to a police station, and attacked a commercial bank. In the ensuing clash with JTF, nine were reportedly killed.

3.2.4 Wudil

During a JAS uprising in July 2009, Wudil LGA was among those that experienced violence with an attack on a police station by insurgents. In November 2012, unidentified gunmen killed a policeman and a member of a local vigilante group.

3.2.5 Doguwa

Few incidents were reported in Doguwa LGA during this period. However, in 2012, suspected JAS gunmen attacked a small town, killing one. Also in 2012, a gun battle between a local vigilante group and a 20-man gang of robbers results in the death of one gang member and the vice chairman of the vigilante group.

3.3 Jigawa State

Incidents per capita rank 36/37; fatalities per capita 36/37 (see Fig. 3.4).

Located in northern Nigeria and sharing a northern border with The Republic of Niger, Jigawa is the eighteenth largest and eighth most populous state in Nigeria. Its approximately 4.4 million people (2006 census) are predominately of Hausa and Fulani background, with Kanuri found in the Hadejia Emirate and some people of

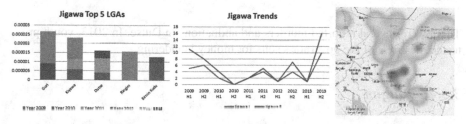

Fig. 3.4 Bar Chart shows annual per capita incidents for top LGAs; Line Graph shows trend in # of incidents and fatalities in the state; Heat Map shows hotspots of violence from 2009 to 2013— Nigeria Watch data formatted and uploaded to Peace Map (www.p4p-nigerdelta.org/peace-building-map)

Badawa background in the northeastern parts of the state. Eighty percent of the population is found in the rural areas of the state, principally engaged in subsistence farming and animal husbandry as well as in a variety of small and medium scale trade and commerce activities in the informal economy, especially in agricultural goods, livestock, and consumer goods. In 2001, the Federal Office of Statistics classified Jagawa State as among those with relatively high incidence and severity of poverty, with a gross annual per capita income of N35,000 (US$290), below the national average. The first civilian governor elected in the Fourth Republic was Ibrahim Saminu Turaki (ANPP). He was succeeded in 2007 by Sule Lamido (PDP).

On a per capita basis, Jigawa was Nigeria's second-least violent state between 2009 and 2013. The few incidents that were reported mainly related to petty crime, clashes between farmers and herdsmen, and limited spillover of the conflict between Boko Haram and the JTF into the state. The following LGA-level breakdown describes risk factors in the five most violence-prone LGAs in the state. The incidents described are not exhaustive but give a brief overview of the scope and tenor of the types of issues reported in the data during the five year period of 2009–2013.

3.3.1   Guri

The most often reported conflict risk factor in Guri was inter-communal violence between pastoralists and farmers. Such incidents in 2012 and 2013 resulted in several fatalities. In 2013 there was also a clash between JTF and suspected insurgents, which killed several.

3.3.2   Kiyawa

Incidents in Kiyawa related to armed robbery and a lynching of two suspected cattle thieves in 2013. Adding pressure to the administration of the state was significant flooding in 2012, Nigeria's most severe rainy season in decades.

3.3.3   Dutse

There were several reported murders and abductions in Dutse. In 2013, youth protested the arrest of the governor's sons for money laundering.

3.3.4   Ringim

Adding stress to the administration of the state serious flooding was reported in August 2012. In 2013, several clashes between security forces and robbers killed about eight people. In one such incident in April an armed gang reportedly bombed a bank and attacked a police station.

3.3.5 Birnin Kudu

There were two incidents reported in Birnin Kudu of clashes between farmers and pastoralists resulting in fatalities (2009, 2012). A couple of murders were also reported in 2009.

3.4 Katsina State

Incidents per capita rank 35/37; fatalities per capita 34/37 (see Fig. 3.5).

Katsina state is located in northern Nigeria, bordering Zamfara, Kaduna, Kano, and Jigawa states, as well the Republic of Niger to the north. Its approximately 5.8 million people (2006 census) are predominately of Hausa and Fulani background. Agriculture is the foundation of the state's economy, with approximately 75 % of its populations involved in farming activities. Crops include guinea corn, millet, maize, cow peas, cotton, and groundnut. Agriculture is supplemented by forestry, livestock rearing, as well as the extraction of minerals, including kaolin, asbestos, gold, uranium, nickel, chromite, silica sand, and clay (www.katsinastate-lgac.com/history.html).

On a per capita basis, Katsina was one of Nigeria's least-violent states between 2009 and 2013. Most of the violence in the state is due to general petty crime without any greater unifying thread. In addition, the state has been the site of repeated deadly floods over the years, most notably in August 2010, July 2011, and September and October 2012. Finally, the state has also seen limited clashes between farmers and herders, as well as the rare spillover of the conflict between Boko Haram and the Nigerian state.

The first civilian governor of Katsina was Umaru Musa Yar'Adua (PDP), elected in 1999. When Yar'Adua was later elected President of Nigeria in 2007, he was succeeded as governor by Ibrahim Shema (PDP). The following LGA-level breakdown describes risk factors in the five most violence-prone LGAs in the state. The incidents described are not exhaustive but give a brief overview of the scope and tenor of the types of issues reported in the data during the five year period of 2009–2013.

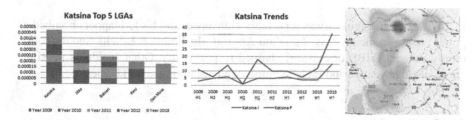

Fig. 3.5 Bar Chart shows annual per capita incidents for top LGAs; Line Graph shows trend in # of incidents and fatalities in the state; Heat Map shows hotspots of violence from 2009 to 2013—Nigeria Watch data formatted and uploaded to Peace Map (www.p4p-nigerdelta.org/peace-building-map)

3.4.1   Katsina

In 2009 police killed a couple members of "Boko Haram," which at that time were called Nigerian Mujahideen, Nigerian Taliban, or Al Sunna Wal Jamma. Also that year, there was a reported case of ritual killing and the lynching of an Army officer who killed a bystander at a mosque in a car accident. The lynching reportedly led to retaliation against the Imam. In 2010, local hoodlums called Kauraye killed a student at the traditional horse riding festival of Eid-Il-Kabir. In 2011, there were protests during the PDP presidential primaries in January and rioting after the election results in April. In 2012, there were peaceful protests against an anti-Islamic video during which French, Israeli, and U.S. flags were reportedly burned. Overall, observers noted a marked increase in security presence in the LGA. In 2013, Kauraye thugs reportedly killed a student during a clash. Other incidents of armed robbery and flooding were also reported throughout the five year period.

3.4.2   Jibia

In Jibia there were reports of ritual killing, two case of inter-communal violence between pastoralists and farmers, and the shooting death of a rice smuggler coming into Katsina from neighboring Republic of Niger by the Nigeria Customs Service.

3.4.3   Batsari

In Batsari LGA inter-communal clashes between pastoralists and farmers were reported. Migration patterns among pastoralists in neighboring Niger raised concerns about the possibility of cross-border inter-communal clashes in 2012. Also reported in Batsari were a number of killings in the course of armed robberies.

3.4.4   Rimi

In 2012 a French engineer was reportedly kidnapped, and his kidnappers bombed a police station in the process. Jama'atu Ansarul Muslimina fi Biladis Sudan (commonly called JAMBS or Ansaru) claimed responsibility. Other incidents in Rimi during the five year period include murder and a shooting at a JTF checkpoint.

3.4.5   Dan Musa

Clashes between vigilantes and cattle rustlers killed several people in 2013. In at least one case it was suspected to have been inter-communal in nature.

Chapter 4
Middle Belt Overview

The Middle Belt region, located in the central part of the country, is culturally diverse with a predominantly Christian population to the south and Muslim to the north. In addition to criminality, political tension and terrorism, conflict emblematic of this region is inter-communal and tends to fall along several overlapping fault lines: (1) farmers versus pastoralists, (2) Christians versus Muslims, and (3) indigenes versus non-indigenes. In Plateau, for instance, there has been significant levels of violence in the northern LGAs between the ethnically Berom (who are predominantly Christian farmers and considered to be indigenes) and the ethnically Fulani (who are predominantly Muslim pastoralists and considered to be non-indigenes). Farther south, in Wase and Langtang local government areas, there has been violence between the ethnically Fulani and the ethnically Tarok (predominantly Christian farmers.) In Benuc and Taraba states there has been violence between the Fulani and the Tiv (also predominantly Christian farmers).

But violence has not always conformed to a certain schematic. In Benue state, for instance, there has also been communal violence between the Tiv and Agatu farmers, while in Nasarawa state, the conflict map has been completely different with violence escalating dramatically in 2012 and 2013 between the ethnically Eggon farmers and a number of different communities, including the Alago, the Fulani, and the Koro. Inconsistencies in the conflict patterns suggest the need for a deeper analysis of the system dynamics in the region, rather than the typical reduction to a simple set of us/them dyads. Such an analysis would take into consideration the complexities of local histories, ethnic and religious polarization, and political and economic conditions at the local, state, regional, and national levels to assess which identity groupings (religious, political, economic, ethnic, etc.) would pull more strongly at any given point in time.

Overall, between the years of 2009–2013, violence was at its most lethal during the Jos riots of 2010. However if that spike in deaths is excluded from regional fatalities, overall numbers are on an upward trajectory, signifying either a rise in overall violence in the region or improved incidence reporting. Historically, Jos has experience several periods of extreme violence. Major crises occurred in 2001,

© Springer International Publishing Switzerland 2015
P. Taft, N. Haken, *Violence in Nigeria*, Terrorism, Security,
and Computation, DOI 10.1007/978-3-319-14935-6_4

2008, and 2010 giving the region a reputation of instability. Election cycles are also times of high tension for Plateau state and throughout the region. At times insurgents have staged attacks exacerbating existing political, communal, and sectarian tensions, particularly since 2011. Organizations like Search for Common Ground have implemented conflict resolution programs in Plateau state, specifically for communities surrounding Jos, by using instruments including collaborative dialogues, community interventions, and peace radio programming. The data included in this section also highlight how states of emergency implemented by the central government may have had an influence on reducing the impact of serious violence in some areas even as it escalates elsewhere.

Due to the complexities of local histories, ethnic and religious polarization, and political and economic conditions, a single explanation for understanding violence across the Middle Belt is infeasible. Collective religious, political, economic, or ethnic identities pull more strongly in different places at different times, producing different environments of conflict within the states. Human geography is also important as urban areas such as Jos in Plateau may experience quicker escalations of violence, while rural areas are more prone to back-and-forth raids between herders and farmers. However, the fault lines of ethnicity and other powerful identifiers that bind communities together clearly play a significant role in defining the conflict landscape (see Fig. 4.1).

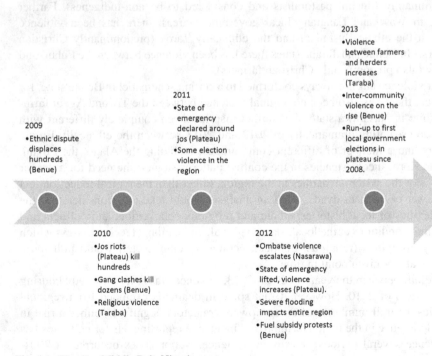

Fig. 4.1 Timeline Middle Belt, Nigeria

4.1 Plateau State

Incidents Per Capita Rank 4/37; Fatalities Per Capita 2/37 (see Fig. 4.2).

Plateau state stretches along Nigeria's "Middle Belt" region in the center of the country. Its estimated population of 3.2 million has over 40 ethno-linguistic groups. The dominant occupations are subsistence farming and pastoral grazing, although mining is also carried out in the tin rich state. Historically, Plateau has been a tourist and retreat destination due to the pleasant weather and scenic vistas. However, inter-communal violence has escalated in recent years. In 2008, local government election violence led to the deaths of hundreds. In 2010, hundreds more were killed after ethnic riots broke out in the capital city of Jos. Much of the violence in Plateau is between those considered to be indigene (many of whom are Christian farmers) and those considered settlers (many of whom are Muslim pastoralists). While the lethality of the violence has come down since the Jos riots of 2010, the number of incidents has been increasing over the last two and a half years. President Jonathan imposed a state of emergency from December 2011 to June 2012 in the four Local Government Areas (LGAs) of Jos North, Jos South, Barkin Ladi, and Riyom. While most of the conflict risk factors have generally been reported in that particular cluster of LGAs, in 2013, there was an escalation of violence reported in the southern LGAs of Wase and Langtang South. In February 2014, Plateau had its first local elections since 2008 when violence killed hundreds. Plateau has been governed by the PDP since the beginning of the Fourth Republic in 1999. Jonah David Jang, a former military governor of Benue and Gongola, was elected governor of Plateau in 2007 and reelected in 2011 (www.plateaustate.gov.ng). The following LGA-level breakdown describes risk factors in the five most violence-prone LGAs in the state. The incidents described are not exhaustive but give a brief overview of the scope and tenor of the types of issues reported during the five year period of 2009–2013.

4.1.1 Riyom

On a per-capita basis Riyom has significant conflict risk, with frequent communal violence between Fulani pastoralists and Berom farmers. Riyom was one of the four LGAs clustered around the city of Jos that was under a state of emergency from December 2011 to June 2012.

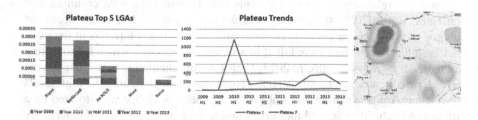

Fig. 4.2 Bar Chart shows annual per capita incidents for top LGAs; Line Graph shows trend in # of incidents and fatalities in the state; Heat Map shows hotspots of violence from 2009 to 2013—Nigeria Watch data formatted and uploaded to Peace Map (www.p4p-nigerdelta.org/peace-building-map)

In 2010 there were numerous reports of inter-communal violence. In March, for instance it was reported that a suspected Fulani militia attacked a Berom village, killing about ten people. Similar incidents were reported throughout 2011 (January, April, August, September, October, and December). Of the five periods, 2012 was the most violent as measured by the number of incidents. In March 2012, two anti-riot policemen were reportedly killed by a suspected Fulani militia. In May several more villages were reportedly attacked. In June, unidentified gunmen attacked and killed John Daring, PDP chairman at Sharubutu Ward, as well as his wife and child, in the village of Rim. Inter-communal conflict continued into 2013, with fewer attacks than in 2012, but often more lethal, especially in March.

4.1.2   Barkin Ladi

Massive levels of sectarian violence in Jos, in March 2010, spread to Barkin Ladi with a number of attacks resulting in fatalities, frequently between the ethnically Fulani and the Berom. Such violence continued into 2011. After Jonathan imposed the state of emergency, the levels of violence seemed to come down during the first half of 2012. But then after it was lifted in June, data from Nigeria Watch, Council on Foreign Relations, and ACLED all suggest that violence in Barkin Ladi spiked considerably, when gunmen reportedly attacked villages on numerous occasions. In July 2012, violence reportedly killed over 100 people during a series of attacks, including one on a funeral in which two prominent politicians were killed. Later in the year, two pubs were reportedly attacked, killing several. In October 2013, about 20 people were reportedly killed in a cattle rustling incident. Then in December about 40 Berom villagers were reportedly killed in an inter-communal attack.

4.1.3   Jos East/North/South

In Jos East, Jos North, and Jos South LGAs an estimated 1,082 people were reportedly killed in 99 incidents between 2009 and 2013. Due to the high population of the area, on a per capita basis, the tally of incidents is lower than some of the other LGAs, although it is still the third most violent LGA in the state out of 17. In 2009, violence was relatively low, with incidents mainly associated with criminals and vigilantes. But in 2010, sectarian riots broke out in Jos, leading to the deaths of hundreds. Ethnic, sectarian, and criminal violence continued throughout the year. On Christmas Eve, dozens were killed in a series of bombings, which led to rioting and reprisals by angry youths. In 2011, Hausa youths attempting to register to vote were reportedly attacked by Christians. People marched, protesting the high levels of violence. A police officer was stabbed in February. A village was stormed by a militia in March. After several more riots and militia attacks, a state of emergency was declared in December 2011. In February, March, and June 2012, three churches were reportedly bombed, killing dozens and leading to mob lynchings in reprisal attacks. There was a reported attack on the Police Staff College by gunmen. In July of 2012 there was a reported attack on an Islamic school in Jos South. Other incidents included inter-communal violence and criminality. There was also a protest by a labor union in November 2012 that turned violent. Inter-communal and criminal violence continued into 2013.

4.1.4 Wase

The southern LGAs of Wase and Langtang South have traditionally been relatively peaceful, with most of Plateau's violence occurring in the northern LGAs clustered around Jos. However, in 2013 there was a significant spike in inter-communal violence in Wase between the pastoralist Fulani and the agrarian Tarok people which reportedly killed over 100. In Langtang South there were additional attacks by pastoralists on villages apparently triggered by allegations of cattle theft. Violence in Langtang South during this period reportedly killed over 50 people. Severe flooding exacerbated pressures in the southern part of Plateau State in August 2012.

4.1.5 Bassa

Bassa was also affected by the sectarian violence of January, 2010. Inter-communal attacks and reprisals were reported throughout the year. In March 2012, a police station was reportedly burned down by rioters. In August 2012, the bombing of a police station was reported.

4.2 Nasarawa State

Incidents Per Capita Rank 9/37; Fatalities Per Capita 5/37 (see Fig. 4.3).

Located in the heart of the country, close to the Nigerian capital, the Nasarawa state was created in October 1996 from parts of Plateau state. The state borders Kaduna state in the north, the Abuja Federal Capital Territory in the west, Kogi and Benue states in

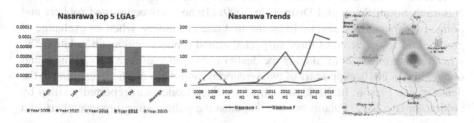

Fig. 4.3 Bar Chart shows annual per capita incidents for top LGAs; Line Graph shows trend in # of incidents and fatalities in the state; Heat Map shows hotspots of violence from 2009 to 2013—Nigeria Watch data formatted and uploaded to Peace Map (www.p4p-nigerdelta.org/peace-building-map)

the south, and Taraba and Plateau states in the east. Nasarawa comprises 13 Local Government Areas, and has a population of a little less than two million people. Twenty-nine different languages are spoken in Nasarawa, including Agatu, Basa and Eggon. Nasarawa's economy relies mostly on agriculture, notably the culture of cash crops such as yam, cassava, and egusi (a variety of melon). Another important part of the state's economy is the production of minerals, such as bauxite and salt. Nasarawa produces a large share of the salt distributed throughout the country. Since the beginning of the Fourth Republic in 1999, governors of Nasarawa were from the ruling PDP, until 2011 when Umaru Tanko Al-Makura won the governorship while running on the Congress for Progressive Change (CPC) platform (www.nasarawastatenigeria.com).

If the fault-lines of violence in the neighboring states have tended to fall more cleanly along religious, indigene/setter, pastoralist/farmer divides, violence in Nasarawa breaks the pattern, especially with the escalation involving the Eggon Ombatse militia[1] in 2012 and 2013. To be sure, disputes over the status of indigeneity and non-indigeneity has played a role in the conflict landscape. But the Ombatse have clashed with both indigenes and non-indigenes, Christians and Muslims, farmers and pastoralists. In 2012, there were clashes between Eggon and Alago, Eggon and Fulani, Eggon and Koro, Eggon and security forces. In January–February, 2013, the Ombatse clashed again with Fulani. Then in May the most lethal Ombatse incident occurred of the five-year period when security forces reportedly attempted to raid the Ombatse shrine, leading to a clash which killed dozens. Some analysts consider the militia to be motivated by their traditional religious ideology. Others see the violence through more of a political lens, particularly in regards to the run-up to the 2015 election. The following LGA-level breakdown describes risk factors in the five most violence-prone LGAs in the state. The incidents described are not exhaustive but give a brief overview of the scope and tenor of the types of issues reported in the data during the five year period of 2009–2013.

4.2.1 Keffi

While Keffi had the highest number of incidents reported per capita (it has one of the smallest populations in the state), it was among the lowest in terms of fatalities per capita during the period of 2009–2013, because the incidents of violence reported in Keffi were generally small scale acts of criminality as compared with the more lethal group-based violence reported elsewhere in the state, especially around Nasarawa South Senatorial District. In 2009, clashes between armed robbers and police officers claimed several lives, the first one in January, when an exchange of gunfire lead to the death of three suspected armed robbers, and again in late 2009, when another gun duel resulted in the death of one police officer, two civilians and one robber. In 2010, supporters of two senatorial candidates clashed, resulting in the death of one person. In 2012, the son of a village leader was reportedly killed by cultists. In late 2012, Fulanis protested against the police, after three members of a

---

[1] The Ombatse object to their classification as a militia, emphasizing their connection to the Azhili shrine and the goal of promoting social good and morality.

family were reportedly killed by local police. In 2013, there was a case of domestic violence in which a woman killed her daughter-in-law. Four were killed during a student protest over the scarcity of water and power in their institutions. In June, a gunman was reportedly killed when a gang attacked the chairman of the Code of Conduct Tribunal's convoy. A young boy of only 10 years old was reportedly beheaded in August, for ritual purposes.

4.2.2 Lafia

Lafia, the capital of Nasarawa, was at the heart of the Ombatse violence in 2012 and 2013. Until that time, however, violence was quite low. In 2009 there were a couple of youth protests. In 2010 there was an incident of gang violence in which two students were killed, an abduction, and a political rally that turned violent. In 2011, an election year, political tensions ran high. There were a number of rallies and protests that turned violent. In February, the CPC candidate for governor, Al-Makura, was briefly detained. His detention led to violent protests which reportedly led to the deaths of four. Al-Makura won the election and became governor in 2011. In October 2011 there was a report of a clash between police and Muslim worshippers, when police disrupted their Friday prayers on the way to the police station. The incident did not lead to any casualties.

Then in the summer of 2012, the Ombatse issue began to emerge with a clash over land and crop payments between the Eggon and the Alago that reportedly killed dozens of people. In November, there were clashes between suspected Ombatse and security forces as well as an attack by suspected Ombatse on a Koro community. In 2013, clashes between Eggon and Fulani reportedly killed dozens. In May, security forces were reportedly on their way to raid an Ombatse shrine when they were ambushed in an incident that reportedly killed dozens near Lafia. This was followed by renewed violence between the Eggon and Alago.

Separately, in 2012 there was also a prison protest and an attempted jailbreak that reportedly killed one person. Two women were also assaulted in the incident. Floods from the seasonal rains killed several. Hundreds of youth protested the inadequacies of the infrastructure development, drainage systems, and disaster preparedness which they felt contributed to the deaths of the young people. A child, accused of witchcraft, was lynched. In 2013 there was an incident of gang violence, a clash reported between two transport unions, a protest by members of the National Youth Service Corps, and a youth protest that resulted in the temporary closure of the state assembly.

4.2.3 Keana

Keana, which is located near the border of Benue state, experienced a surge in intercommunal violence during the five-year period. In 2011, a clash between Tiv farmers and Fulani herdsmen killed several people. In 2012 it was reported that more clashes broke out between Tiv and Fulani after refugee camps were closed. Inter-communal violence continued into March, killing dozens. In October, a clash between Fulani and Eggon was reported. In 2013, several incidents of inter-communal violence reportedly killed about 18 people.

4.2.4   Obi

In 2009, one youth was reported killed in a chieftaincy tussle in an Alago town. There were no reported incidents until 2012, when inter-communal violence (Fulani/Koro and Eggon/Koro) killed dozens. Inter-communal violence continued into 2013 with clashes between Fulani and Tiv, Fulani and Eggon, as well as Eggon and Alago, which killed dozens.

4.2.5   Akwanga

Akwanga was generally peaceful until 2012 when unknown gunmen reportedly raided Fulani herdsmen, killing over a dozen. In 2013 a Tiv community was reportedly attacked. In September a clash was reported between security forces and suspected Ombatse, which led to the deaths of about a dozen people.

4.3   Benue State

Incidents Per Capita Rank 14/37; Fatalities Per Capita 7/37 (see Fig. 4.4).

Benue state is comprised of 23 Local Government Areas (LGAs) and named for the Benue River, which flows from northern Cameroon through Adamawa, Taraba, and Benue, and Kogi where it meets the Niger River. Benue's total population is estimated to be about five million people with most residents speaking either Idoma or Tiv. Makurdi, the capital of Benue, has about 600,000 residents and resides in a rich agricultural region. Potatoes, cassava, yams, and flax are a few of the crops produced there. The state's economy is based largely on agricultural output with 80 % of the population engaging in food production. This makes Benue and its residents particularly vulnerable to natural disasters. Benue's current governor, Gabriel Suswam was elected in 2007. He is a member of the People's Democratic Party (PDP), as was his predecessor (www.ngex.com/nigeria/places/states/benue.htm).

Instances of violence and insecurity spiked in the election year of 2011 and then escalated again dramatically in 2013. The LGAs most affected by insecurity were those closest to the capital city and the Nasarawa border. In 2009–2012 the main causes of violence in Benue were criminal and political. But in 2013 inter-communal clashes and sectarian violence increased significantly, displacing thousands.

Fig. 4.4 Bar Chart shows annual per capita incidents for top LGAs; Line Graph shows trend in # of incidents and fatalities in the state; Heat Map shows hotspots of violence from 2009 to 2013—Nigeria Watch data formatted and uploaded to Peace Map (www.p4p-nigerdelta.org/peace-building-map)

The following LGA-level breakdown describes risk factors in the five most violence-prone LGAs in the state. The incidents described are not exhaustive but give a brief overview of the scope and tenor of the types of issues reported in the data during the five-year period of 2009–2013.

4.3.1   Makurdi

In 2009 there were a few reports of fatalities associated with crime and domestic violence. In two separate incidents, an official with the Benue State Ministry of Lands and Survey and the chairman of the Trade Union Congress were killed by gunmen. A clash between Hausa Fulani and Tiv reportedly killed about eight, over a dispute regarding the use of land and water by pastoralists. In 2010 rival cult groups (gangs), reportedly clashed in January and November, killing about a dozen. In the November incident, Black Axe and Red Axe cult groups reportedly clashed, leading to the deaths of seven. In 2011 thousands were displaced by inter-communal violence and reprisal attacks between Fulani pastoralists and Tiv farmers (February, June, November). Several murders were also reported, including the killing of a member of the Action Congress of Nigeria (ACN), and the Special Advisor to the Benue State Governor on Media and Public Affairs. In 2012 there were protests over the removal of the fuel subsidy, student protests over campus crime and police brutality, an incident of intra-communal violence over land in a Tiv community, and a clash between two cult groups. Also reported was a clash between police and Muslim youth over the planned demolition of property including a mosque. There were several murders reported, and flooding of the Benue River in the most destructive rainy season in decades. In 2013 inter-communal violence escalated dramatically with incidents of violence between Tiv and Fulani (April, May, and November) and between Tiv and Agatu communities (June). Also reported in the year was a protest by teachers and the National Labour Congress over compensation, a protest by youth over alleged police brutality, and several murders. The Academic Staff Union of Universities (ASUU) peacefully protested for better infrastructural development at educational institutions.

4.3.2   Guma

In 2009, inter-communal conflict between Fulani and Tiv displaced hundreds. In 2011, ballot boxes were stolen in a clash that reportedly killed five. Also in 2011 inter-communal violence killed dozens, and displaced thousands. Inter-communal violence continued into 2012, killing dozens more. In 2013, violence between Fulani and Tiv escalated, with fatalities reported in March, April, May, June, July, August, and December. Other incidents included the torturing of a man accused of witchcraft and a kidnapping.

4.3.3   Agatu

No incidents were reported in Agatu from 2009 to 2012. But in 2013, inter-communal violence escalated dramatically. In May there was a report of an attack by Fulani on a funeral, killing dozens of Idoma. More clashes were reported in July, September, October, and November.

4.3.4   Gwer East/West

As in other Benue LGAs, inter-communal violence was the main issue reported. There were no reports in 2009. In 2010, there was a clash between Hausa and Tiv communities in May. In June a land dispute between two communities reportedly killed eight and displaced hundreds. In 2011, clashes between Fulani and Tiv reportedly killed ten and displaced thousands. Inter-communal violence continued into 2012 with clashes reported in March, April, and May. An attack by political thugs was also reported in March, in which the home of a local chairman was burned down. In 2013, inter-communal violence was reported in March which killed several people including three police officers. Also reported were several fatalities associated with criminal violence.

4.3.5   Apa

Incidents in Apa were mainly interpersonal and criminal as opposed to group-based. A man was reportedly murdered by two sisters in 2009. A member of the Nigerian Security and Civil Defense Corps (NSCDC) was killed by unknown gunmen in 2013. Also in 2013, an 18 year-old woman were murdered and buried on a farm.

Fig. 4.5 Bar Chart shows annual per capita incidents for top LGAs; Line Graph shows trend in # of incidents and fatalities in the state; Heat Map shows hotspots of violence from 2009 to 2013—Nigeria Watch data formatted and uploaded to Peace Map (www.p4p-nigerdelta.org/peace-building-map)

4.4 Taraba State

Incidents Per Capita Rank 31/37; Fatalities Per Capita 20/37 (see Fig. 4.5).

Bordering the Republic of Cameroon, Taraba is a large state geographically (third largest) but sparsely populated. Its population is ethnically diverse, comprising Jenjo, Kuteb Chamba, Mumuyes, Mambila, Wurkums, Fulani, Jukun, Ichen, Tiv, Kaka, Hausa, and Ndoro people. Most are involved in agriculture, animal husbandry, or small scale cottage industries. Products include coffee, tea, groundnuts, cotton, maize, rice, sorghum, millet, cassava, and yam. Leather goods, pottery, metalwork, and dyed cloth are also a significant part of the state's economy. Since the Fourth Republic began in 1999, the governor of Taraba has been a member of the ruling PDP. Danbaba Danfulani Suntai was elected in 2007 and was reelected in 2011 (www.tarabastate.gov.ng).

Taraba was the most peaceful state in the Middle Belt region during five year period, although in 2012 and 2013 there were spikes in violence principally related to clashes between farmers and herdsmen. At times there was a sectarian dimension, with religion serving as identifier and justification for acts of violence. The following LGA-level breakdown describes risk factors in the five most violence-prone LGAs in the state. The incidents described are not exhaustive but give a brief overview of the scope and tenor of the types of issues reported in the data during the five year period of 2009–2013.

4.4.1 Jalingo

Although violence in Taraba overall has been relatively low over all between 2009 and 2013 compared to other Nigeria states, the number of incidents reported in Jalingo LGA increased in the 2011 election year with a political rally that turned violent, and a number of murders. A policeman was killed in December by unidentified gunmen. In 2012 the police commissioner was targeted by suicide bombers on motorcycles, killing about ten. The following month an IED exploded near a bank. In May there was a reported youth protest over the ban of motorcycle taxis. In June, gunmen reportedly shot a police officer. A bomb reportedly exploded in a pub in October. In 2013, it was reported that there was inter-communal violence between Tiv and Kutep.

4.4.2   Takum

In 2009 there were reports of clashes between the Jukuns and Kutep. There were no reports of violence in 2010. In 2011, reported inter-communal clashes between Tiv and Fulani killed several. In 2012 multiple clashes between Fulani and Tiv (February, March, and April) reportedly killed dozens and destroyed villages. There were allegations that some of the perpetrators were recruited from outside the area. In some of these incidents security forces were also killed and their uniforms stolen. Separately, an Irish construction worker was killed by unknown gunman. In 2013, an inter-communal clash was reported between Tiv and Kutep.

4.4.3   Ibi

In 2012, violence between Christians and Muslims escalated. In one incident a "Christian vigilante group" reportedly killed a Muslim, triggering clashes which killed about 10. In 2013, several incidents of inter-communal violence reportedly killed about a dozen.

4.4.4   Ardo-Kola

In 2011 there was a clash reported between people of two different ethnic groups over a land dispute, killing several. In 2013, armed robbers reportedly killed two at a cattle market.

4.4.5   Wukari

In 2010 there was reportedly a clash and a series of retaliation attacks between Christians and Muslims sparked by a disagreement over the proposed location of a mosque, leading to the deaths of about a dozen people. In 2012, a gang of about 30 gunmen attacked several banks, a police station, and a pub killing several. Flooding displaced thousands. In 2013, an argument between two youths sparked a sectarian clash, reportedly killing between 5 and 31 people. In 2013, inter-communal violence with sectarian overtones broke out when herdsmen reportedly attacked a Jukun funeral, killing about 30 people.

Chapter 5
Northeast Overview

In the international media, Boko Haram and #BringBackOurGirls are emblematic of conflict and insecurity in Nigeria. But this is reductive both to Nigeria as a whole and to conflict in the Northeast itself. As detailed in this book, across the country there are patterns of criminal, intra-communal, inter-communal, ethno-sectarian, political, and separatist conflict drivers and trends. Emanating from the Northeast, the phenomenon of Boko Haram has elements of all those types. Notwithstanding the fact that Boko Haram was designated as a terrorist group by the U.S. Department of State in November 2013, there is no group that calls itself Boko Haram. Before 2009, such extremists were sometimes called the Nigerian Taliban or the Nigerian Mujahideen. Residents of Maiduguri, in Borno State, eventually started calling them Boko Haram, which is a Hausa derivation referring to something ambiguous (Books? Education? Fraud? (Murphy 2014)) as being religiously forbidden, or haram. Usually, when people said Boko Haram, they were referring to the followers of Muhammed Yusuf, whose own group was actually called *Jamā'at Ahl as-Sunnah lid-da'wa wal-Jihād* (JAS). Now, every time a bomb explodes, a police officer is shot, a bank is robbed, or a village is attacked, the incident is attributed generically in the media to Boko Haram, regardless of whether JAS had anything to do with it. The various objectives of violence seem inconsistent and at times even at cross-purpose. While many attacks are no doubt inspired by ideology, others are ethno-sectarian, or criminal. Some attacks have more to do with state or national politics than any radical jihadist agenda, which would seek to usurp or replace the existing political structures. The anarchic nature of the violence suggests a more complex dynamic than a simple diagnosis of insurgency.

That said, insurgency certainly plays a big part. JAS was founded by Yusuf in 2002, and gradually radicalized until the uprising of July 2009 (Sergie and Johnson 2014), when they attacked a number of targets in Bauchi, including a police station to steal weapons. Hundreds were killed in the incident. Yusuf was captured in Maiduguri two days later and died in police custody on July 30, 2009. After his death, the insurgency died down. In late 2010, there was a massive prison break in Bauchi, releasing many who had been arrested after the 2009 uprising. Then, under

© Springer International Publishing Switzerland 2015
P. Taft, N. Haken, *Violence in Nigeria*, Terrorism, Security,
and Computation, DOI 10.1007/978-3-319-14935-6_5

the new leadership of Abubakar Shekau, the JAS insurgency began to escalate rapidly, particularly in Borno state. Meanwhile, and perhaps adding fuel to the insurgency, 2011 was an election year in Nigeria. When President Goodluck Jonathan, from the Niger Delta state of Bayelsa, was declared the winner, post-election violence erupted in many northern states, including Borno, Yobe, Adamawa, Bauchi, and Gombe states. JAS surely sought to capitalize on the widespread public sentiments of political discontent and stoke sectarian tension in an attempt to further destabilize the Northeast.

Meanwhile, in 2012, another group *Jamā'atu Anṣāril Muslimīna fī Bilādis Sūdān* (Ansaru or JAMBS), emerged in reaction to the more indiscriminate violence perpetrated by JAS, focusing their attacks mainly on Western targets and security installations. While JAS was operationally centered mainly on Borno State, JAMBS appears to have operated farther west, including from Kano and Katsina.

Following renewed violence in late 2010, the government has struggled to both anticipate and respond to the attacks against both its own forces as well as the civilian population. According to a June 2014 report by the U.S. Congressional Research Service, multiple factors hamper the ability of security forces, including both military and police, to effectively deal with threats emanating from terrorist/insurgency groups. First and foremost is a continued lack of cooperation and coordination between various security entities, both at the national and local levels. Corruption, inadequate resourcing and training, and a lack of morale among soldiers, particularly from impoverished areas in the Northeast, also compound the problem. Local communities, in self-protection, have often resorted to forming their own vigilante groups to repel incursions, including networks of informants who try to act in an early warning capacity. In both cases, attacks against both government troops and local vigilante groups have been increasingly deadly, with the families of soldiers as well as informants being targeted for reprisal.

As regards trends, violence continued to escalate throughout the region until May 2013, when President Jonathan declared a state of emergency in Borno, Adamawa, and Yobe states. A heat map query of the Peace Building Map prior to the state of emergency shows the vast majority of violent incidents took place near the city of Maiduguri. Upon implantation of the state of emergency, violence reduced dramatically in Maiduguri, only to be displaced to the more rural LGAs where it escalated even more dramatically (http://library.fundforpeace.org/cungr1420). Attacks became increasingly lethal. The numbers of fatalities soared. The type of violence also changed in the wake of the state of emergency. If violence previously targeted key leaders, state assets, and infrastructure, after the state of emergency there were many more reported attacks on minority villages in rural areas, burning homes and churches, killing dozens at a time in what has appeared in some cases to have the hallmarks of ethnic cleansing.

Other than the May declaration of a state of emergency, government-led counter-insurgency operations have been in the form of military interventions, including the bombardments of training camps, and joint military-police invasions. As noted above, since May 2013 the federal and local governments have also sanctioned the operation of local vigilante forces, generally referred to as the Civilian Joint Task Force (CJTF). The CJTF often operated in conjunction with or alongside government and military forces during the state of emergency and has been heralded with cautious optimism by experts both inside and outside of Nigeria. To some, it merely underscores the inadequacy of the state and local security services in countering the threat of terrorist and insurgent groups. To others, it is a demonstration of local ownership of the problem and a community-led effort to expel extremist elements from the region (Campbell 2013).

In addition to the soaring fatalities and the loss of property, the conflict in the region has also caused an enormous amount of displacement. Although the exact number is not clear, a September 2014 report by the European Commission estimated that at least 70,000 people have fled to Niger, Chad and Cameroon and between 650,000 and 3.3 million people have been internally displaced (European Commission, Humanitarian Aid and Civil Protection 2014).

Overall, the states in the region where the highest intensity of violence occurred during the time period analyzed were Adamawa, Borno and Yobe. Borno state has been the most conflict-affected. While Adamawa has had relatively fewer incidents than the other two states, its shared border with Cameron makes it a security priority for the government, particularly as insurgents are believed to use the area as a crossing point as well as a potential staging ground. Borno is also the second largest state in Nigeria and has a history of political instability and high levels of violence. Since its independence, the local government has changed 14 times, including seven military coups. Between May 2011 and August 2013, out of 8,600 deaths, caused by the political, economic and social violence observed nationwide by the Council on Foreign Relations' Nigeria Security Tracker, 2,470 (28.7 %) occurred in Borno (Action on Armed Violence 2014). The capital of the state, Maiduguri, has been the most violent LGA in the state since the beginning of the escalation of the conflict in 2011. As noted above, after the declaration of the state of emergency in May 2013, the violence in Borno has started to spread to the rural LGAs.

Another state of high insecurity, Yobe has one of the largest cattle markets in West Africa and has also suffered from sporadic outbreaks of violence between militants and security services. In July 2013, schools across the state closed because of an attack at a secondary school, killing 42 students. Like Yobe, Adamawa is of high strategic importance since it connects Borno to other states, which has made it vulnerable to attacks as well as a refugee center for populations fleeing from Borno. The state was deeply affected by flooding in July 2012, causing displacement, property damage and fatalities, mainly in Yola and Lemurde. Meanwhile, Bauchi state has also experienced sporadic insurgent violence and has the unfortunate distinctions of having one of the highest poverty rates in the country, estimated at 49 %, as well as the second highest unemployment rate, at an estimated 30 % (Action on Armed Violence 2014).

On the opposite end of the spectrum, one of the least violent states of the region is Gombe. The state has a relatively smaller land area of 20,265 km^2 and a highly diverse population. Conflict resulting in fatalities spiked in 2012, particularly in the capital city of Gombe, where incursions by suspected insurgents led to attacks on churches and police stations. Other incidents, while occasionally reported as the work of terrorists or insurgents, may well have been merely criminal, with attacks on banks and police stations occurring with the most frequency (see Fig. 5.1).

Timeline

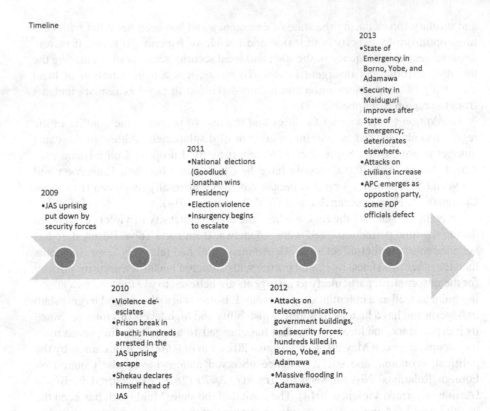

2013
•State of
Emergency in
Borno, Yobe, and
Adamawa
•Security in
Maiduguri
improves after
State of
Emergency;
deteriorates
elsewhere.
•Attacks on
civilians increase
•APC emerges as
opposition party,
some PDP
officials defect

2011
•National elections
(Goodluck
Jonathan wins
Presidency
•Election violence
•Insurgency begins
to escalate

2009
•JAS uprising
put down by
security forces

2010
•Violence de-
esclates
•Prison break in
Bauchi; hundreds
arrested in the
JAS uprising
escape
•Shekau declares
himself head of
JAS

2012
•Attacks on
telecommunications,
government buildings,
and security forces;
hundreds killed in
Borno, Yobe, and
Adamawa
•Massive flooding in
Adamawa.

Fig. 5.1 Timeline Northeast, Nigeria

5.1   Borno State

Incidents Per Capita Rank 1/37; Fatalities Per Capita Rank 1/37 (see Fig. 5.2).

The second largest state in Nigeria, Borno is home to approximately 4.2 million people (2006 census) of primarily Kanuri, Bura and a few nomadic Shuwa Arab ethnicities. The state is predominantly Muslim. The second-largest producer of maize in Nigeria, Borno also produces millet, rice, wheat, and cotton. The state is strategically positioned as a transit point for consumer goods to and from the neighboring countries of Niger, Chad and Cameroon. The All Nigeria People's Party (ANPP)—previously called the APP—has held the governorship of the state since the beginning of the Fourth Republic in 1999. Governor Ali Modu Sheriff held the office from 2003 to 2011 and was succeeded by Governor Kashim Shettima (www.bornostate.gov.ng).

The state of Borno, nicknamed "The Home of Peace," is not historically violent. But in July 2009, clashes between militants and security forces broke out in several northern states with about 700 reportedly killed in Borno's capital city of Maiduguri.

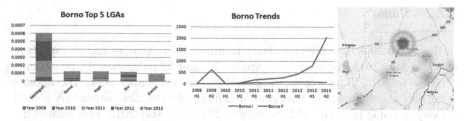

Fig. 5.2 Bar Chart shows annual per capita incidents for top LGAs; Line Graph shows trend in # of incidents and fatalities in the state; Heat Map shows hotspots of violence from 2009 to 2013—Nigeria Watch data formatted and uploaded to Peace Map (www.p4p-nigerdelta.org/peace-building-map)

Violence initially came back down, but then in 2011 it began to escalate, quickly becoming the most violent state in the country. Attacks on churches, schools, government buildings and officials have killed and displaced thousands. In May 2013, President Goodluck Jonathan imposed a state of emergency in Borno, Yobe, and Adamawa calling for "extraordinary measures to restore normalcy." The latter half of 2013 has consequently been deadly for suspected militants, with hundreds losing their lives to air strikes, military raids and armed clashes with youth vigilante groups such as the Borno Vigilance Youth Groups (BVYG). The following LGA-level breakdown describes risk factors in the five most violence-prone LGAs in the state. The incidents described are not exhaustive but give a brief overview of the scope and tenor of the types of issues reported in the data during the five year period of 2009–2013.

5.1.1 Maiduguri

The capital and largest city in Borno, Maiduguri has been the most violent LGA in the state since the insurgency began escalating in 2011. Militants have planted IEDs, bombed markets, attacked police stations, raided military bases, stormed schools and killed a handful of prominent individuals (e.g. district heads, state officials, councilors, politicians, clerics, university professors). Under the state of emergency, declared in 2013, the city has been put on overnight curfew, and 12 areas in the city have been under permanent curfew. The JTF has raided several insurgent strongholds in the city, killing hundreds and discovering numerous arms caches. Clashes between militants and youth vigilante groups have also broken out around town, often leading to fatalities and reprisal attacks. According to data cross validated across Nigeria Watch, ACLED, and the Nigeria Security Tracker, the number of incidents of violence dropped precipitously in Maiduguri after the declaration of the state of emergency. However, violence in rural LGAs increased, surpassing Maiduguri for the first time in the summer of 2013.

5.1.2 Bama

In Bama, militants have raided military facilities, police stations and prisons. One attack in May 2013, reportedly claimed the lives of 22 policemen, 14 prison officers, two soldiers, 13 insurgents, three children and one woman. Police checkpoints at the border with Cameroon have also sustained violent attacks, particularly around the city of Banki in April 2012, November 2012, and May 2013. In response, the JTF and vigilante groups have raided many of the training camps located in the region (March 2012, May 2012, January 2013 and July 2013).

5.1.3 Kaga

In May, 2013, two clerics were reportedly killed. Several attacks were reported in July 2013 on the town of Mainok by Boko Haram insurgents, killing 23 civilians. Attacks on police stations and LGA officials were reported as well. In September 2013, clashes were reported between militants and vigilante groups. In October, 2013, a military offensive reportedly killed dozens.

5.1.4 Biu

Biu was the fourth most violent LGA on a per capita basis, where insurgents have frequently targeted religious communities. In 2011 a Muslim cleric was murdered in his home. In June 2012, a church was reportedly attacked during Sunday mass, killing two and injuring hundreds. In July 2013, 13 were handcuffed and burnt to death in their church. In August 2012, there were attacks at mosques killing two individuals. A year later, eight people, including teachers and clerics were killed in an attack. Raids by the JTF and youth vigilante groups were carried out in response. Other attacks included a shooting of three men playing cards in 2012 and several other clashes and protests. In August 2013, several people were reportedly killed in a clash between youth vigilante groups.

5.1.5 Gwoza

Violence in Gwoza LGA, on the border of Cameroon, escalated in 2013 after the state of emergency was declared. Villages inhabited by minorities were targeted by suspected insurgents, killing worshipers, and torching churches and houses. The Nigerian military raided militant strongholds in May 2013 and employed airstrikes in November.

5.1.6   Other LGAs

5.1.6.1   Konduga

In Konduga LGA there were attacks on telecommunication towers in 2012, as well as multiple assassinations of clerics, village heads and traditional rulers. In response, the Nigerian military, with assistance from youth vigilantes, raided multiple militant camps in May, June and July 2013, killing insurgents and recovering arms. After May 2013, there was the reported use of air strikes on suspected insurgent camps.

5.1.6.2   Damboa

In Damboa LGA, militants killed the former chairman of the LGA in February 2012, a district head and the secretary of the assembly in February 2013, and a village head in April 2013. Insurgents also attacked telecommunication offices and government buildings in September–November 2012. Some military raids were reported in January, July and August 2013. In October 2013, militants reportedly shot several people at a mosque. In November, militants attacked several villages, killing dozens. Vigilante groups retaliated.

5.1.6.3   Kukawa

In Kukawa LGA, a huge clash between Boko Haram and the JTF reportedly left close to 200 dead, with 2,000 homes, 62 cars and 284 motorcycles destroyed in April 2013.

5.2   Yobe State

Incidents Per Capita Rank 11/37; Fatalities Per Capita Rank 4/37 (see Fig. 5.3).

Yobe state has an estimated population of 2.3 million people, who are mainly from the Kanuri ethnic group. An agricultural state, Yobe produces acacia gum, groundnuts, beans, cotton and has one of the largest cattle markets in West Africa, located in Potiskum. Also found in the state are fishing grounds and mineral deposits such as gypsum, kaolin and quartz. As in Borno state, the All Nigeria People's Party (ANPP) has held the governorship since the beginning of the Fourth Republic in 1999.

Yobe was one of the northern states (along with Borno and Bauchi) that experienced a spate of violence between Islamist militants and security services in July 2009, particularly in Potiskum and Nangere LGAs. Such insurgent violence receded after that initial spike, then began to escalate rapidly in 2011. Although violence had begun to come back down in 2013, President Goodluck Jonathan imposed a state of

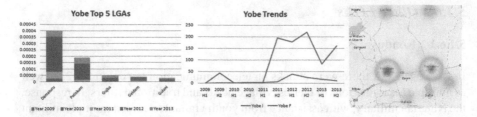

Fig. 5.3 Bar Chart shows annual per capita incidents for top LGAs; line graph shows trend in # of incidents and fatalities in the state; Heat Map shows hotspots of violence from 2009 to 2013—Nigeria Watch data formatted and uploaded to Peace Map (www.p4p-nigerdelta.org/peace-building-map)

emergency in Yobe, Adamawa and Borno in May of that year. In July 2013, Governor Ibrahim Gaidam (All Nigeria People's Party) ordered the closure of schools across the state, following an attack at a secondary school in Mamudo in Ngere LGA, which left 42 students dead. It is estimated that about 20,000 residents of North Eastern states have been displaced by violence and in November 2013 the president appealed to legislators to extend the state of emergency for another six months. The following LGA-level breakdown describes risk factor in the five most violence-prone LGAs in the state. The incidents described are not exhaustive but give a brief overview of the scope and tenor of the types of issues reported in the data during the five year period of 2009–2013.

5.2.1 Damaturu

In 2010 there were a couple attacks reported on police and on a lawmaker. In April, 2011 there were election protests after Goodluck Jonathan was declared the winner. Later in the year, large scale attacks by insurgents targeted government buildings, banks and churches in bombings and shootings, killing at least 63. In December, a gun battle between militants and security forces reportedly led to the deaths of over 50 people. Churches were bombed on Christmas. In early 2012 it was reported that Habib Bama, a sect leader suspected to be behind an attack on a church on Christmas day 2011, had died from gunshot wounds inflicted by soldiers attached to the Joint Task Force during a raid. With a heavy presence of security forces in the LGA, there were some reports of violence against civilians. Insurgents warned southerners to leave the state, then reportedly bombed a pub. Others, including a cleric and Chadian nationals were also killed during the year. In June 2012 there was reportedly unrest and violence, killing 52. In August, a suicide bomber attacked a military check-point, killing eight. In both 2012 and 2013 there were jailbreaks reported, releasing dozens of suspected insurgents. After a school massacre by militants in September 2013 killed 42 students, security forces conducted airstrikes on a camp. Militants responded with a coordinated attack on military barracks and police in October, which reportedly killed over 100 people.

5.2.2 Potiskum

Potiskum, the commercial hub of the state and headquarters of the Fika and Potiskum Emirates reported high levels of violence. The most common incidents related to clashes between security forces and gunmen and attacks on police stations, mosques and churches. A few incidents of criminal violence were also reported. In July 2009, fighting between militants and security forces killed dozens. Violence didn't escalate again until 2011, when there was reportedly a cattle raid, sectarian mob violence, and a gun battle reported between militants and security forces, which caused hundreds to flee in early January 2012. Attacks on a pub, a church, and on traders appeared to target southerners. In separate incidences a customs official, a vigilante leader, a military intelligence officer, a retired colonel, and a cleric were also killed in 2012. Police stations were attacked. In May, 2012 a cattle market was attacked in an incident that led to the death of about 50, reportedly in retaliation for the lynching of a suspected cattle thief. Others alleged that the attack was carried out by militants. In June, the Deputy Majority Leader of the Yobe State House of Assembly was reportedly killed by gunmen. In August, the Emir of Fika was reportedly targeted by a suicide bomber at the Potiskum mosque, in an attack which left six people severely injured. In October a school was reportedly burned by militants. A police station, a bank, and a church were attacked in December. In 2013, three Korean doctors were killed and one of them beheaded in their home by unknown assailants. A security official and two clerics were also reportedly killed in 2013. In July dozens were reportedly killed in an attack on a boarding school. A village was attacked in October, killing four.

5.2.3 Gujba

In 2012, a school, a court, police stations, and churches were reportedly attacked. In 2013 there were reports of clashes between security forces and militants. In a September incident, shops, houses, and telecommunications facilities were destroyed, killing as many as 100 people. Also in September, about 50 students were killed in an attack on a school. In response to the school attack, airstrikes reportedly killed dozens suspected militants.

5.2.4 Geidam

Geidam, located on the border of the neighboring country of Niger was reportedly the fourth most violent LGA in the five year period, although violence did not escalate in Geidam until late 2011 when militants reportedly attacked a pub, a police station, a bank, some churches, and a few other buildings. In 2012, a few southern traders were reportedly killed. Following a series of threats on his life, the district head of Geidam, was killed in February 2012. In December an attack on a village reportedly killed about ten. In 2013 a few clashes between security forces and militants were reported.

5.2.5 Gulani

In 2010 a police station was reportedly attacked in Gulani. In 2013 it was reported that there were attacks on a police station and a telecommunications facility. Explosives were stolen from a construction site in an attack by suspected militants in July.

5.3 Adamawa State

Incidents per Capita Rank 24/37; Fatalities per Capita 12/37 (see Fig. 5.4).

Formed in 1991, the northeastern state of Adamawa is one of the largest states in Nigeria. It borders the country of Cameroon to the east, Borno state to the north, Gombe state to the west, and Taraba state to the south. Its location makes it a key corridor between Borno, a hub of insurgent activity, and other states. Its population of about 3.5 million mainly comprises farmers and cattle herders. The economy is predominately agricultural, although the state also has some mineral wealth. Common crops include maize, millet, sorghum, rice, yams, and cassava. Cotton and groundnuts are also produced as cash crops. Unlike Borno and Yobe, Adamawa has traditionally been governed by the ruling People's Democratic Party (PDP), although in November 2013, Governor Murtala Nyako defected to the newly formed All Progressives Congress (APC) along with the governors of Rivers, Sokoto, Kano, and Kwara states (www.adamawa.gov.ng).

On May 16, 2013, President Goodluck Jonathan declared a state of emergency in Borno, Yobe, and Adamawa, because of spikes in violence and terrorist activity in

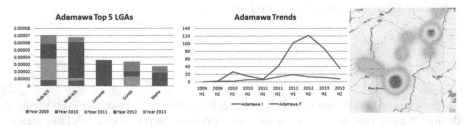

Fig. 5.4 Bar Chart shows annual per capita incidents for top LGAs; Line Graph shows trend in # of incidents and fatalities in the state; Heat Map shows hotspots of violence from 2009 to 2013—Nigeria Watch data formatted and uploaded to Peace Map (www.p4p-nigerdelta.org/peace-building-map)

the area attributed mainly to Boko Haram. Although in Adamawa, by the time the state of emergency was declared, violence had already begun to come down. Of the three states, the one with the most severe levels of violence was Borno to the north, causing thousands of refugees to flee south into Adamawa. While Borno was the most violent state in the country as measured by incidents and fatalities per capita over the five year period, Adamawa was the 12th most violent as measured by fatalities per capita and 24th out of 37 as measured by incidents per capita. The following LGA-level breakdown describes risk factors in the five most violence-prone LGAs in the state. The incidents described are not exhaustive but give a brief overview of the scope and tenor of the types of issues reported in the data during the five year period of 2009–2013.

5.3.1 Yola North/South

Issues around the town of Yola included violence along sectarian lines as well as terrorist attacks. In January 2011 a riot broke out at a prison during an attempted jailbreak. In April 2011 there was rioting and protests after the Presidential elections. Protesters who had been detained staged a riot in the jail where they were being held. A number of killings associated with criminal violence were also reported during the year. In 2012 there were a number of attacks on churches, pubs, and southern traders. Other shootings were attributed to armed robbers and other unidentified groups, including an attack in May 2012 that reportedly left ten people dead. Adding further stress to the administration of the state was severe flooding in the worst seasonal rains in decades after water was released from a dam across the border in Cameroon, causing temporary displacement and upheaval. In 2013 a number of clashes between security forces and unidentified gunmen were reported. In February 2013, 11 people reportedly died in sectarian fighting. A police station was reportedly burned down in September.

5.3.2 Mubi North/South

As in Yola, riots broke out after the 2011 elections. In January 2012, gunmen reportedly attacked a meeting of southern traders, killing 20. In October 2012, dozens of students were killed in attacks believe to be connected to the student union elections at the Federal Polytechnic. Others suspected that insurgents may have been behind the attack. In October and November 2012, explosions in Mubi reportedly targeted Joint Task Force (JTF) patrols, killing at least 22 people. In September 2013, it was reported that youth vigilantes apprehended several suspected insurgents who later died from injuries sustained at the hands of the youth. In December a currency exchange was reportedly attacked by gunmen. In the clash that followed over a dozen were reportedly killed.

5.3.3 Lamurde

In January and May of 2012, inter-communal violence killed up to 50 people. The May incident was reported to be a reprisal attack by herdsmen in response to a previous clash in January. In addition to the killings, there was looting and burning of farm produce and homes. Animals were also reportedly stolen.

In October, following Friday prayer, IEDs went off that were purportedly targeting the JTF, who were also involved in a shoot-out with unknown gunmen during the month. As in Yola, August flooding caused wide-scale displacement and property damage when Cameroon opened a dam across the border. Although many parts of Adamawa experienced negative effects from the flooding, Larmude appeared to be particularly hard hit, with almost a dozen deaths attributed to the disaster.

5.3.4 Gombi

In 2011, suspected insurgents robbed a bank, and attacked two police stations, reportedly killing up to 20 people. Separately, in December there was a clash reported between security forces and an armed gang. In 2012 there were several reported deaths suspected to have been associated with the counter-insurgency operation after an IED exploded a few days previously. In August 2013, a man reportedly killed his son for ritual purposes. In September, gunmen reportedly attacked a police station, killing two.

5.3.5 Maiha

Maiha LGA is located on the border of Cameroon. Cross-border banditry is reported to be a concern. In August 2010, it was reported that robbers from Cameroon killed ten. In December 2012, militants thought to be associated with Boko Haram attacked local government buildings including a police station, killing a reported 30 and freeing at least 35 prisoners. In May 2013 an attack on a church and village market reportedly left at least ten dead in the village of Jilang in Maiha. In a statement given after the attack, a police spokesperson noted that the assailants had likely utilized illegal routes into the LGA from neighboring Cameroon, allowing them to also flee and avoid capture. In a similar statement following the December attack on the police station, the porous border with Cameroon was once again mentioned as problematic in allowing assailants to sneak into and out of the LGA with little to no detection.

5.3.6 Other LGAs

5.3.6.1 Madagali

As in the other LGAs, Madagali also suffered throughout 2012 from attacks blamed on individuals and groups believed to be associated with Boko Haram. In particular, 2012 saw at least three attacks on local police stations, with one being burned to the ground in December. In April 2013, an attack on the village of Midlu left at least 11 dead and several others injured in an assault on the deputy governor's home and surrounding residences. While initial reports blamed the attacks on Boko Haram, others attribute the attack on lingering political tensions between the deputy governor and House speaker based on the last local elections in which each were vying for the PDP nomination, with the speaker eventually securing the needed votes. During the attack, the deputy governor's daughter was also reported kidnapped, although she was later released unharmed.

5.3.6.2 Ganye

Although Ganye LGA reported fewer incidents than others, it merits some discussion as one of the few LGAs to show an increase in violence in 2013. In February 2013, pastoralists allegedly raided a farm, killing one man. Local farmers stormed the pastoralist camp in retaliation, killing ten. Although 2012 was mainly characterized by reported violent clashes between pastoralists and local settlers, incidents of violent attacks, shootings and killings attributed to insurgency/counter-insurgency also increased throughout the first half of 2013. In March, suspected militants stormed a town attacking a police station, a bank, and causing a jail break. About 30 fatalities were reported in all. During this attack, reports cited the deliberate targeting of civilians along with police and JTF personnel.

5.3.6.3 Hong

Hong LGA deserves mention as throughout 2012, militants were very active in the area and attacked several churches, opened fire at a student housing residence, and planted bombs targeting security forces. In January 2012, gunmen shot 17 people at a funeral apparently along ethno-religious lines. In October, alleged militants attacked a student housing area, killing 48 students and injuring at least 15 others. In June, bombs were discovered at a police station and a filling station but were dismantled by a bomb squad. In late 2012, the JTF reportedly discovered an insurgent bomb-making facility while over 100 suspects were arrested, mostly in connection with the October attacks.

5.4   Bauchi State

Incidents Per Capita Rank 23/37; Fatalities Per Capita Rank 10/37 (see Fig. 5.5).

Bauchi state borders Yobe state in the northeast, Gombe state in the east, Taraba state in the southeast, Plateau state in the south, Kaduna state in the west, Kano state in the northwest, and Jigawa state in the north. Nicknamed the Pearl of Tourism, the state of Bauchi was created in 1976, after the former North Eastern State was broken into several smaller states. It used to include Gombe state, which later became its own state in 1996. The state got its nickname, the Pearl of Tourism, thanks to the Yankari Game reserve and the Rock Paintings at Goji and Shira. Consisting of 20 Local Government Areas (LGA), the state occupies a total land area of 49,119 km². Its landscape is mostly mountainous and rocky in the western and northern regions of the state because of its proximity to the Cameroon Line and to the Jos Plateau. Bauchi's primary agricultural products include maize, millet, groundnut and corn, and livestock. Demographically, the state comprises a total of 55 distinct tribal and linguistic groups, including Hausa, Fulani, Kanuri, Zulawa and Badawa. The first governor of Bauchi state in Nigeria's Fourth Republic was elected in 1999 as a member of the PDP. In 2007, Isa Yaguda, a member of the ANPP, succeeded him. In 2009, Yaguda switched to the PDP and was reelected in 2011 (www.bauchistate.gov.ng).

Bauchi was at the epicenter of the so-called "Boko Haram Uprising" in July 2009, when Islamist militants attacked a police station. Over the next few days, violence spread to Borno and Yobe. All told, over 1,000 people were reportedly killed in the space of less than a week across the region. Since then, in Bauchi, the levels of violence have been less, however in 2011, over 100 people were reportedly killed around the time of the election. The following LGA-level breakdown describes risk factors in the five most violence-prone LGAs in the state. The incidents described are not exhaustive but give a brief overview of the scope and tenor of the types of issues reported in the data during the five year period of 2009–2013.

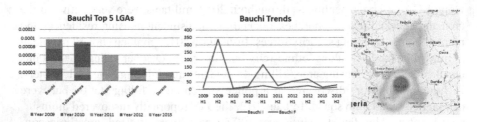

Fig. 5.5 Bar Chart shows annual per capita incidents for top LGAs; Line Graph shows trend in # of incidents and fatalities in the state; Heat Map shows hotspots of violence from 2009 to 2013—Nigeria Watch data formatted and uploaded to Peace Map (www.p4p-nigerdelta.org/peace-building-map)

5.4.1 Bauchi

In February 2009, a disagreement over worship space between Christians and Muslims reportedly led to rioting and the deaths of about 12 people. Three months later, in May, students rioted over alleged irregularities in the administration of their exams. Two were reportedly killed by police. On July 26, Islamists attacked a number of targets in Bauchi, including a police station to steal weapons. Hundreds were killed. In December 2009, a religious sect, called Kala Kato confronted police after being prohibited from open-air preaching. Over 30 people were reportedly killed and hundreds displaced in the incident. In 2010, dozens were reportedly killed in religious protests and protests by students. In the summer of 2010, it was reported that there were clashes political clashes over the upcoming elections to be held in April 2011. Two political aspirants were reportedly attacked. In September 2010, hundreds escaped in a prison break, including many who had been arrested for the Islamist violence of July 2009. In October 2010 two policemen and a community leader were reportedly killed by gunmen in separate incidents. Also in 2010 an Islamic scholar and a lecturer were reportedly killed. In April 2011, political violence escalated. When the results were announced, rioters reportedly killed about 30 people and burned down dozens of churches. On the day of Jonathan's inauguration, about 12 were killed in a bomb blast in Bauchi. In the second half of 2011 there were a number of murders reported and at least one roadside bomb, which caused several injuries. In 2012 protests broke out across the country over the government's removal of the fuel subsidies. A suicide bomb at a church reportedly killed about a dozen people. Several IEDs exploded in the summer of 2012. Several policemen were killed by unidentified gunmen. Telecommunications facilities were reportedly attacked. In September a suicide bomber reportedly attacked a Catholic church, killing several. Also in 2012 there were a few reports of violence perpetrated by Sara-Suka gangs. In 2013, an insurgent group called Jama'atu Ansarul Muslimina Fi Biladis-Sudan (JAMBS), commonly known as Ansaru, reportedly kidnapped and killed several foreigners. Several other protests and murders were also reported during the year.

5.4.2 Tafawa-Balewa

In 2010 a Christian pastor and his wife were reportedly kidnapped and killed. A rally protesting President Goodluck Jonathan led to one death and several injuries. When several people were arrested for vandalizing a community radio station, a mob attacked the police station. One person was killed in the incident. In 2011, a policeman was reportedly killed during voter registration. A dispute at a billiards game reportedly led to an outbreak of violence between Christians and Muslims in which several people were killed, mosques and houses destroyed, and thousands reportedly displaced. Inter-communal violence involving pastoralists reportedly killed several in March of 2011. In 2012 a predominantly Christian village was attacked, in an incident that reportedly killed three. Several were reportedly killed in a bank robbery. There were several reports of inter-communal violence, some-times between Hausa-Fulani and the Sayawa communities.

5.4.3 Bogoro

Many Christians of the Sayawa ethnic group live in Bogoro LGA. In 2011 several inter-communal attacks on villages in Bogoro were reported, in which dozens were killed and property destroyed.

5.4.4 Katagum

In 2010 a suspected rapist was lynched. In 2011, the main issue reported in Katagum was insurgent activity against security forces. In 2012, a number of prominent indi-viduals, including the former Comptroller General of Prisons were reportedly killed. After gunmen killed a member of the Katagum Emirate Council, hundreds of youths apprehended the suspected perpetrators. In the course of the incident, several people were reportedly killed.

5.4.5 Darazo

Attacks on banks and police stations were reported in Darazo in 2011 and 2013.

5.5 Gombe State

Incidents Per Capita Rank 32/37; Fatalities Per Capita Rank 27/37 (see Fig. 5.6).

Gombe borders Yobe state to the north, Borno state to the east, Adamawa state to the southeast, Taraba state to the south and Bauchi state to the west. It was formed in October 1996 by the Abacha military government with parts of the Bauchi state. Nicknamed the Jewel in the Savannah, the state comprises 11 Local Government Areas (LGA), and has a land area of 20,265 km^2. The local population is made up of a large number of different ethnicities, including Fulani, Tangale, and Waja. The land-scape is mountainous in the southeastern part of the state, and flat savannah plains in the rest of the state. Gombe is a major food basket for the nation, producing cereals, legumes, fruits such as lemons, mangos and guava, tree crops, fish and livestock.

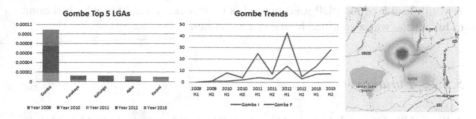

Fig. 5.6   Bar Chart shows annual per capita incidents for top LGAs; Line Graph shows trend in # of incidents and fatalities in the state; Heat Map shows hotspots of violence from 2009 to 2013—Nigeria Watch data formatted and uploaded to Peace Map (www.p4p-nigerdelta.org/peace-building-map)

At the beginning of the Fourth Republic in 1999, an ANPP governor was elected. In 2003, the governorship switched to the PDP with the election of Abubakar Mohammed. Mohammad Danjuma Goje was elected governor in 2011, also a member of the ruling PDP party.

Although Gombe was among the least violent states during the five-year period, there were a few spikes in incidents resulting in fatalities, particularly in 2012. The following LGA-level breakdown describes risk factors in the five most violence-prone LGAs in the state. The incidents described are not exhaustive but give a brief overview of the scope and tenor of the types of issues reported in the data during the five year period of 2009–2013.

5.5.1   Gombe

In April 2011, following the announcement of the results for the presidential elections which brought Goodluck Jonathan to power, the Gombe LGA, as the capital of the Gombe state, experienced some violence. Violent protests by supporters of the CPC political party (Congress for Progressive Change), upset about their party's loss in the election, including an attack on the chairman of the PDP (People's Democratic Party), led to the deaths of five people. Political tensions exacerbated polarization along ethnic and religious lines. In 2012, 15 incidents were reported in the Gombe LGA, leading to a total of 38 fatalities. In January 2012 there were several attacks by suspected insurgents, who opened fire on Christian worshipers inside a church, killing six people and injuring ten more. A few weeks later, gunmen raided a pub in the city of Gombe. In February 2012, suspected insurgents attacked a police station after having failed to storm a jail, in order to free some of the inmates. Others killed in 2012 included a cleric, a traditional ruler, and the Government House Acting Director of Security Matters. In 2013, a lower amount of incidents were reported, but the Gombe LGA still saw bursts of violence, including when gunmen on motorcycles opened fire on people playing cards, which resulted in the death of five people and injured seven more.

A youth leader of the PDP was shot at his residence. A neighbor who had come to his rescue was also killed by the same gunmen.

5.5.2 Funakaye

A bomb killed three children in May 2011. The bomb reportedly had been left by armed robbers who had raided a bank the day before, during which three people were killed. In March 2012, an attack on banks and a police station by suspected insurgents, left four policemen and three civilian dead.

5.5.3 Kaltungo

Incidents in 2010 reported in Kaltungo LGA related police action against a gang of criminals and a clash between two communities that led to the death of ten people. Three hundred persons reportedly took refuge in a school after their houses and farms had been destroyed. In 2013, another communal clash was reported that led to the death of two.

5.5.4 Akko

In 2010 a clash was reported during a gubernatorial political rally that killed two. In 2012, suspected insurgents reportedly attacked two churches, a beer parlor, and two police stations. Unidentified gunmen attacked the home of a military commander. Police disarmed a bomb found at Gombe State University. In 2013 there were several attacks on public institutions, including the raiding of the Kumo divisional police station and of the military base, which killed one police officer. Also in 2013, an attack on the headquarters of the Akko local government in Kumo reportedly resulted in the death of one policeman and one civilian. In late 2013, two men conspired to kill another one over a conflict involving a girlfriend.

5.5.5 Kwami

In October 2011, a bomb attack on a mobile police base by suspected insurgents led to the death of one policeman and three attackers, and destroyed one building and 15 vehicles. In late 2013 suspected insurgents attacked a police station. Five were reportedly killed in the incident.

Chapter 6
Northwest Overview

While insurgency raged 300 miles to the east, tallying up the highest numbers of fatalities per capita in the country during the period of 2009–2013, in the Northwest things were relatively calm during this period, with some states (Sokoto and Kebbi) having very few incidents per capita and others, such as the Federal Capital Territory (FCT), with periodic spikes in violence. The FCT had a significant number of incidents and fatalities resulting from bombings, protests, and gang violence. In Zamfara there were a number of highly lethal raids on villages by large gangs and several cases of pastoral/farmer clashes. The conflict patterns in the five states included in this region vary by trend and issue. According to the National Bureau of Statistics, the region is also very diverse socioeconomically, comprising some of Nigeria's poorest states (Zamfara, Kebbi, and Sokoto) as well as Niger, which is one of the wealthier states and the Federal Capital Territory, which is second only to Lagos in terms of having the lowest levels of poverty. While difficult to identify a common thread or theme in this region, disaggregating to the state and LGA-level highlights hotspots and issues of concern (Fig. 6.1).

6.1 Zamfara State

Incidents per capita rank 33/37; fatalities per capita rank 18/37 (see Fig. 6.2).

Zamfara state is located in the far northwestern part of the country states, sharing borders with the states of Sokoto to the northwest, Kebbi to the west, Niger to the south, Kaduna to the southeast, Katsina to the east, and the Republic of Niger to the north. Created in 1996, the state is composed of 12 LGAs. Demographically, Zamfara state mostly comprises Hausa and Fulani, with some other minority ethnic groups such as Zamfarawa, Gobirawa and Burmawa.

Agriculture is the main source of income, the state's slogan being "farming is our pride." With a landmass of about 39,000 km², agriculture employs more than 80 % of the population. The main agricultural products are millet, maize, cotton and beans. The state also possesses significant solid mineral deposits, including gold,

© Springer International Publishing Switzerland 2015
P. Taft, N. Haken, *Violence in Nigeria*, Terrorism, Security,
and Computation, DOI 10.1007/978-3-319-14935-6_6

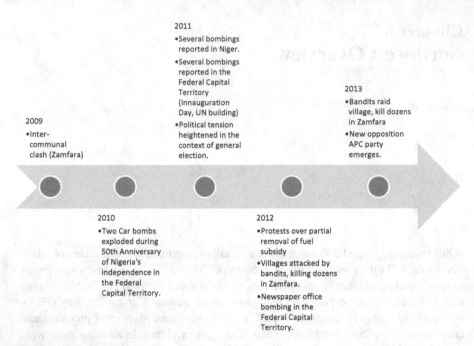

Fig. 6.1 Timeline Northwest, Nigeria

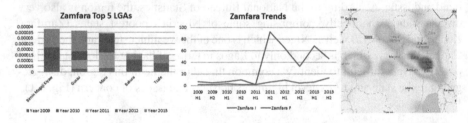

Fig. 6.2 Bar Chart shows annual per capita incidents for top LGAs; Line Graph shows trend in # of incidents and fatalities in the state; Heat Map shows hotspots of violence from 2009 to 2013—Nigeria Watch data formatted and uploaded to Peace Map (www.p4p-nigerdelta.org/peace-building-map)

copper, and iron. Tourist attractions include the ancient settlement of Jata. According to the event data, Zamafara has a low frequency of violent incidents, though some of those incidents have been quite lethal, especially after 2011. Politically, the state has been governed by a member of the ANPP since the beginning of the Fourth Republic in 1999 (www.ngex.com/nigeria/places/states/zamfara.htm). The following LGA-level breakdown describes risk factors in the five most violence-prone LGAs in the state. The incidents described are not exhaustive but give a brief overview of the scope and tenor of the types of issues reported in the data during the five year period of 2009–2013.

6.1.1 Birnin Magaji/Kiyaw

Unlike the northeastern states, where large scale attacks have been tied to Islamist insurgency, and the Middle Belt where they have often been connected to inter-communal violence, attacks on villages by gangs in Birnin Magali/Kiyaw have mostly been associated with criminality and armed robbery, or reprisal attacks between criminal gangs and vigilante groups. In 2012, a dozen villagers were reported to have been killed in one such incident. In January 2013, as many as ten were reportedly killed. Other incidents included a clash in October 2013, between the followers of Darika Islam, and the followers of Izala Islam. Three people, including a ward councilor, were reportedly killed in the incident and a mosque was burned down. Also that month, youths staged a protest against the acting governor for the worsening state of security in the area.

6.1.2 Gusau

As the capital of Zamfara State, Gusau has the largest population of all the LGAs. In 2009 and 2010, there were a few political clashes were reported, including one in August 2010 between PDP and ANPP supporters that reportedly killed as many as ten people. In 2011, gunmen reportedly killed a police officer guarding voter registration equipment. A clash between pastoralists and farmers reportedly killed two. In 2012 a number of murders, sometimes connected to armed robbery, were reported. The former Commissioner of Agriculture was reportedly murdered. Separately, a spike in lead poisoning connected to unsafe mining practices was reported during this time. In 2013, in addition to reports of armed robbery, kidnapping, and clashes between criminals and police, there was also a protest by civil servants over unpaid salaries, and another by youths in a state-sponsored skills acquisition program over unpaid allowances.

6.1.3 Maru

Multiple large scale incidents of armed bandits with motorcycles and AK47s attacking villages or clashing with vigilante groups were reported, killing dozens in 2011, 2012, and 2013. In one such incident in June 2012, about 80 bandits reportedly attacked several villages and killed as many as 23, in reprisal for the killing of bandits by vigilantes the previous year.

6.1.4 Bakura

Several murders were reported in Bakura, including that of a prominent businessman and an Islamic cleric in 2012. In 2013, a hajj registration official was reportedly killed by gunmen.

6.1.5   Tsafe

Although incidents were few during this period, some of them were quite severe, especially in 2013. In 2009, there was an inter-communal clash reported between herdsmen and a farming community which killed several. In 2010, gunmen reportedly attacked the ZSIEC (Zamfara State Independent Election Commission) chairman's residence, killing two. In 2013 it was reported that a large number of gunmen raided a village and killed dozens.

6.2   Sokoto State

Incidents per capita rank 34/37; fatalities per capita rank 35/37 (see Fig. 6.3).

Sokoto state is located in the northwestern part of Nigeria, sharing borders with the states of Zamfara in the south, Kebbi in the southwest, and the Republic of Niger in the north. Created in 1976, the state has a landmass of about 28,000 km², divided into 23 LGAs.

Demographically, the state is mainly populated by Hausa and Fulanis. Minority groups include the Zabarmawa, Tuareg and Dakarkari. Most of the inhabitants are Muslim, with a large majority of them being Sunni, and a minority Shia. Most are farmers, producing cash crops such as cotton and rice, as well as millet and cassava. Land is also widely used for grazing cattle and goats. Sokoto exports goats, sheepskins and finished leather products. Mineral resources include limestone and kaolin. Tourist attractions include the Surame Cultural Landscape, the Akalawa Ruins, the Gilbadi Rocks, as well as several monuments, museums and festivals. The first civilian governor in the Fourth Republic, elected in 1999, was Attahiru Bafarawa (ANPP), followed by Aliyu Magatakarda Wamakko (PDP) in 2007 (www.sokotostate.gov.ng). Reports of violence in Sokoto were rare, as compared to other Nigerian states, although there was an increase in 2012 and 2013. The following LGA-level breakdown describes risk factors in the five most violence-prone LGAs in the state. The incidents described are not exhaustive but give a brief overview of the scope and tenor of the types of issues reported in the data during the five year period of 2009–2013.

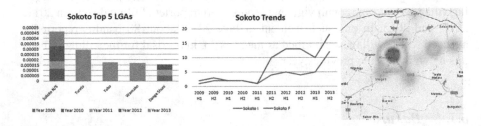

Fig. 6.3 Bar Chart shows annual per capita incidents for top LGAs; Line Graph shows trend in # of incidents and fatalities in the state; Heat Map shows hotspots of violence from 2009 to 2013—Nigeria Watch data formatted and uploaded to Peace Map (www.p4p-nigerdelta.org/peace-building-map)

6.2.1 Sokoto North/South

The area with the highest prevalence of conflict risk factors reported during the five year period was around the state capital in Sokoto North and Sokoto South LGAs. In 2011, a protest reportedly broke out when a man was alleged to have desecrated the Koran. Though Islamist insurgents associated with Boko Haram rarely penetrated as far west as Sokoto State, there were a few incidents attributed to them during this period, including the killing of two kidnapped expatriates and a few attacks on security forces in 2012. Several prominent individuals were reportedly killed, including a police inspector in 2012, and a prison deputy controller and a member of the Sokoto State House of Assembly in 2013.

6.2.2 Tureta

The main issue in Tureta was some violence between armed robbers and police reported in 2013.

6.2.3 Yabo

There were a couple murders reported in Yabo, including the strangling of the UNDP Sokoto Coordinator in 2013.

6.2.4 Wamako

Several incidents related to Islamist insurgency/counter-insurgency were reported in 2013, including the killing or capture of insurgent leaders in March, July, and August. Other issues reported in Wamako included a clash between Sunni and Shia Muslims in 2010, a student protest over scholarship payments in 2013, and a couple murders.

6.2.5 Dange Shuni

Reports in Dange Shuni included the murder of a prominent PDP official in 2011 and a popular Islamic cleric in 2012.

6.3 Kebbi State

Incidents per capita rank 37/37; fatalities per capita rank 37/37 (see Fig. 6.4).

Kebbi was the least violent state in Nigeria, according to the event data, both in terms of incidents and fatalities over the five year period. Located in the northwestern

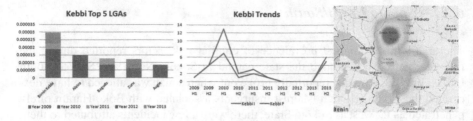

Fig. 6.4 Bar Chart shows annual per capita incidents for top LGAs; Line Graph shows trend in # of incidents and fatalities in the state; Heat Map shows hotspots of violence from 2009 to 2013—Nigeria Watch data formatted and uploaded to Peace Map (www.p4p-nigerdelta.org/peace-building-map)

part of the country, Kebbi state, formed in 1991, is divided into 21 Local Government Areas, and shares borders with the states of Sokoto in the northeast, Zamfara in the east and Niger in the south, as well as with the Republics of Niger and Benin in the west. With a landmass of about 36,000 km², Kebbi state is home to a large number of ethnic groups, mainly Hausa, but also Bussawa, Dukawa, Kambari and Kamuku.

The primary source of livelihood in Kebbi is the cultivation of crops, such as millet, corn, beans, wheat, ginger, and tobacco. Fruit grown in Kebbi includes mangos, guava and pawpaw. Kebbi also has an important fishing industry, in the Niger River which flows through the state. A key tourist attraction is the annual Argungu fishing and cultural festival.

The first civilian governor of Kebbi, Adamu Aliero (ANPP), was elected in 1999. He was succeeded by Usman Saidu Nasamu Dakingari (PDP) in 2007. He was reelected in 2011, but the results were annulled and Aminu Musa Habib Jega was appointed acting governor until a rerun could take place. The following LGA-level breakdown describes risk factors in the five most violence-prone LGAs in the state. The incidents described are not exhaustive but give a brief overview of the scope and tenor of the types of issues reported during the five year period of 2009–2013.

6.3.1   Birnin Kebbi

There were a few murders reported in Birnin Kebbi, including that of a politician in 2009. In the same year, one person was killed in a fight between a farmer and a pastoralist. There were a couple suspected ritualistic killings in 2010. Also in 2010 a couple of politicians were killed by cars in collisions that were suspected to be deliberate. In the election year of 2011 (March) there was a political clash when a mob attacked the convoy of a gubernatorial candidate. In May, two expatriates were reportedly kidnapped by Islamist insurgents. They were later killed in neighboring Sokoto. In 2013, retired teachers protested over pension payments.

6.3.2   Aleiro

In Aleiro, the main issues reported were clashes between police and armed robbers.

6.3.3   Bagudo

In Bagudo, there was a murder and a clash between police and armed robbers. In the election year of 2011 a protest was reported over inadequate electoral materials. One was reported to have been killed in the incident.

6.3.4   Zuru

In 2010 there was a clash reported between the supporters of governor Dakingari and former governor Aliero, although no casualties were reported. Other incidents related to robbery and murder.

6.3.5   Augie

In Augie there were a couple clashes reported between police and armed robbers. In 2012, it was reported that journalists covering a local election were attacked.

6.4   Niger State

Incidents per capita rank 27/37; fatalities per capita rank 26/37 (see Fig. 6.5).

With a landmass of about 76,000 km² and 25 Local Government Areas (LGAs), Niger State is the largest in the country. It contains two of the major hydroelectric power stations of Nigeria. Tourist attractions include the Guara Falls, and the Kanji National Park. The state borders the Republic of Benin to the northwest, and the states of Kaduna, the Federal Capital Territory, Zamfara, Kebbi, Kogi, and Kwara. It is ethnically diverse, with groups such as the Nupe, the Gbagyi and the Hausa. Eighty-five percent of the population is made up of farmers. Minerals resources include talc, gold, silica, sand and marble (www.nigerstate.gov.ng). Relative to the rest of the country, Niger was quite peaceful during this five year period. However, the most conflict prone LGAs in Niger state were Changchaga, Suleja, Tafa, Lapai and Rijau. The year 2011 was the most deadly, with a bombing at the INEC office in Suleja, and several bomb attacks in churches during Christmas

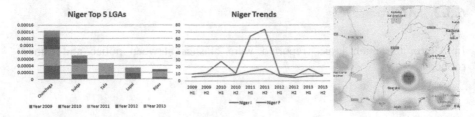

Fig. 6.5 Bar Chart shows annual per capita incidents for top LGAs; Line Graph shows trend in # of incidents and fatalities in the state; Heat Map shows hotspots of violence from 2009 to 2013— Nigeria Watch data formatted and uploaded to Peace Map (www.p4p-nigerdelta.org/peace-building-map)

celebrations. Politically, the state has been governed by members of the PDP since the beginning of the Fourth Republic, in 1999. The following LGA-level breakdown describes risk factors in the five most violence-prone LGAs in the state. The incidents described are not exhaustive but give a brief overview of the scope and tenor of the types of issues reported in the data during the five year period of 2009–2013.

6.4.1 Chanchaga

Issues in Chanchaga LGA included suspected ritual killings of children and a teenager in 2011, 2012, and 2013. There were several protests during the five-year period. In 2010, students protested after a truck killed a student. Another individual was reportedly killed in the protest, when the military was deployed to restore order. In 2011, there were some violent protests by the opposition after the election of President Goodluck Jonathan, in which two persons were reportedly killed. In 2012, there were protests in Chanchaga, as in other parts of the country, over the partial removal of the fuel subsidy, reportedly killing at least three. Also in 2012, commercial motorcycle drivers reportedly protested after a police officer allegedly killed a driver. Suspected Islamist insurgents reportedly killed several police officers in 2012. A woman was reportedly shot by security forces during a counter-terrorism operation. In 2013, two cattle rustlers reportedly murdered a herdsman and stole 45 cows. Also in 2013 a vigilante was reportedly stabbed to death by a "hoodlum." Other reports over the five-year period related to interpersonal violence and criminality.

6.4.2 Suleja

Most incidents reported in Suleja LGA had to do with terrorism and suspected Islamist insurgency in 2011 and early 2012. Bombings were reported in 2011 (PDP rally, school, INEC office, several churches) killing dozens. Also in 2011, gunmen reportedly killed several Igbo traders when they could not recite the

Koran. In 2012 there was another church bombing and the killing of a police officer by suspected Islamist insurgents. Other incidents included several ritual killings in 2009. Two guards to the Suleja LGA chairman were reportedly killed by gunmen in 2011. In 2012, gunmen reportedly raided a police station, killing one. In 2013, a soldier reportedly killed a man during a peaceful Shiite procession after a traffic argument.

6.4.3 Tafa

Incidents in Tafa were mainly criminal or interpersonal.

6.4.4 Lapai

Pastoralist conflict and suspected Islamist insurgency were reported in Lapai over the five year period. In 2010, an incident of inter-communal conflict and reprisal attacks between Fulani pastoralists and a farming community reportedly killed about a dozen people. In 2011 and 2012, suspected Islamist insurgents reportedly killed several police officers. In 2013, there was an incident reported in which a vigilante group shot two herdsmen.

6.4.5 Rijau

Issues in Rijau mainly had to do with gang and election violence. Violence between criminal gangs and police was reported in 2010 and 2011. Also in 2011, a gang of robbers attacked traders seeking to buy animals, killing four. Violence during a local election in 2011 broke out when people accused the electoral staff of partisanship. At least one person was reportedly killed in the incident.

6.5 Federal Capital Territory

Incidents per capita rank 5/37; fatalities per capita rank 9/37 (see Fig. 6.6).

The Center of Unity, as the Federal Capital Territory (FCT) is called, is located in the heart of the country, at the north of the confluence of the Niger River and the Benue River (www.fct.gov.ng). It borders Niger, Kaduna, Nasarawa, and Kogi states. In addition to Nigeria's capital city of Abuja, the FCT comprises five LGAs: Abaji, Gwagwalada, Kuje, Bwari, and Kwali. For the sake of this breakdown we considered the FCT as a whole given the tendency of media reports to refer to Abuja

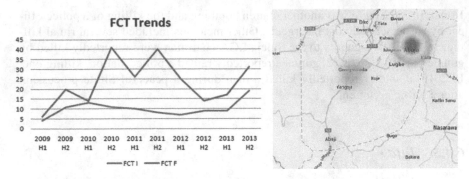

Fig. 6.6 Bar Chart shows annual per capita incidents for top LGAs; Line Graph shows trend in # of incidents and fatalities in the state; Heat Map shows hotspots of violence from 2009 to 2013—Nigeria Watch data formatted and uploaded to Peace Map (www.p4p-nigerdelta.org/peace-building-map)

more generally when talking about the FCT. Sources of income in the FCT include the subsistence farming of crops, such as yam, maize and beans, fishing along the rivers of Usamma, Jabi and Gurara; and wood and craftwork, which has a long tradition in the area. Wood products include masks, musical instruments and tobacco pipes. As the political capital of the country, the city of Abuja is quite diverse, but local ethno-linguistic groups in the region include the Gbabyi, Koro, Gade, Bassa, Gwandara and Ganagana. Ever since the beginning of the Fourth Republic, Nigeria's president has been a member of the PDP, as has been the minister of the FCT.

6.5.1   FTC Summary

Inasmuch as our conflict assessment framework includes both peaceful protests as well as other sorts of violence, and the FCT is the political capital of the country where much political expression takes place, it comes out as one of the more conflict prone localities in the country on a per capita basis during the five year period. In addition to such protests, there were also a number of terrorist attacks during the period, both by Islamist insurgents as well as Niger Delta militants. Other conflict factors included communal clashes and a large number of reports of other forms of criminal and interpersonal violence.

Terrorism and Insurgency: On 1 October 2010, two car bombs exploded during celebrations of the 50th anniversary of Nigeria's independence from Great Britain, killing 12. Henry Okah a MEND leader was arrested in South Africa following the explosions. Three months later on New Year's Eve, another bomb exploded near army barracks, killing four people. Unlike the first bombing, this one was suspected to have been planted by Islamist insurgents, although no one immediately claimed responsibility. A bomb exploded on the inauguration day of the newly reelected President Goodluck Jonathan, killing between zero and four people, depending on the reports. Several were reportedly killed in a suicide bombing at the Abuja police headquarters on June 16. A

JAS spokesperson said the group claimed responsibility. On the outskirts of Abuja, in the neighboring state of Niger, a church bombing killed several people on July 10, 2011. A couple weeks later a bomb at the UN regional headquarters in Abuja killed about 21 people. On 26 April 2012, a bombing at the *This Day* newspaper building in Abuja (and a simultaneous attack at the newspaper's office in Kaduna) killed several people. In June, a bombing at a nightclub reportedly damaged some property. In April 2013, no casualties were reported when a bomb exploded outside a restaurant in Abuja. In June, MEND claimed responsibility for an attack on two fuel trucks in Abuja. In September, security forces raided a building and killed about seven people they thought were Islamist insurgents. An investigation was later launched into whether the victims had actually been squatters.

Political Clashes and Protests. In 2009 one person was reportedly killed in a clash between the Yanbola and the Yangwoza street gangs when one group could not get the other to support their preferred candidate in an upcoming election. PDP supporters clashed with ACN supporters in 2010. Later, ACN demonstrated in protest against attacks by suspected PDP supporters. In March 2010 protestors demonstrated against INEC, in advance of the upcoming 2011 elections. After a local election in April, ACN protested the results. When police intervened to restore order one person was reportedly killed in the confusion. In August, a prominent PDP supporter was reportedly abducted, in what might have been a politically motivated crime. In 2011 some post-election protests broke out on the outskirts of Abuja, reportedly destroying dozens of cars. Shops closed while a heavy security presence was deployed to keep the political violence from escalating. In 2013, tensions within the newly formed APC in Niger State led to a protest at party headquarters in Abuja.

Protests and Labor Strikes. There were a large number of protests, labor strikes, and clashes between labor factions in the Federal Capital Territory during the five year period. In 2010 there were protests by human rights activists, women's groups, a clash between NULGE factions, and a NUPENG strike and electrical workers' strike. In 2011, there were protests over road accidents, a national health bill, and a strike by the state power company workers. Ex-militants from the Niger Delta protested the perceived lapses in the amnesty program. In 2012, as in many other parts of the country, there were protests over the partial removal of the fuel subsidy. There were protests by university students over issues of courses and accreditation, a protest by pensioners, and a protest by Bakassi youths over cross-border problems with Cameroon in the southeastern part of the country. Residents of a condemned building protested the demolition. One person was reportedly killed in that incident. In 2013, the number of protests reported increased. Issues being protested were similar to those of previous years (healthcare, pensions and benefits, minimum wage, demolitions, layoffs, and university matters). Several were arrested in May Day protests. Commercial motorcycle drivers protested against approaches to policing by public security forces. There were some protests against government agencies and officials, including one against the Aviation Minister.

Inter-communal Violence. In 2012 there were reports of clashes between Fulani and Gwari killing two and displacing thousands.

Chapter 7
Southwest Overview

The southwestern region of the country is the heart of "Yorubaland" and hosts Lagos, one of the largest cities in the world. Lagos, which means "lakes" in Portuguese, was the capital of Nigeria from 1914 to 1991, when the seat of government was moved to Abuja. Politically, the southwest has shifted over the course of the Fourth Republic, with a tendency to join cross-regional coalitions in opposition to whichever region claims presidential incumbency when the incumbent is not from the southwestern region. In 1999, when the region's own Olusegun Obasanjo (from Ogun State) came into power on the PDP ticket, the region was solidly AD but shifted to PDP in 2003 when Obasanjo was reelected. After he turned over power to Umaru Musa Yar'Adua (from the North), also of the PDP in 2007, the ACN was ascendant in the region, capturing most state houses in 2010 and 2011. Then, in 2013, with PDP's Goodluck Jonathan from the Niger Delta in the presidency, the ACN joined the CPC, the ANPP, and part of the AGPA to form the All Progressives Congress (APC) in a major reshuffling of the political landscape, laying the foundation for a new cross-regional opposition coalition, including the Southwest and the far North.

While some other regions in Nigeria experienced high levels of communal violence and insurgency during the time range focused on in this book (2009–2013), insecurity in the southwest mainly related to cult violence between groups such as the Eiye, Aye, Lord, Black Axe, and KK confraternities. There were also a number of clashes involving transport unions such as NURTW, RTEAN, and the TOAN, as well as political clashes involving supporters of the PDP and the opposition ACN party. Other issues of violence included domestic violence, crime by "area boys" and the Yoruba nationalist vigilante group, the Oodua People's Congress (OPC). The OPC has historically advocated for the self-determination of the Yoruba people but seems to function primarily as a mechanism for score settling and vigilantism.

Megacities such as Lagos, one of the most important economic centers of the country, pose unique infrastructural, demographic, and governance challenges. As such, Lagos had significantly more incidents than any other state in the Southwest region. The population of Lagos is estimated at 10.7 million (2009), with a growth

© Springer International Publishing Switzerland 2015
P. Taft, N. Haken, *Violence in Nigeria*, Terrorism, Security,
and Computation, DOI 10.1007/978-3-319-14935-6_7

Timeline

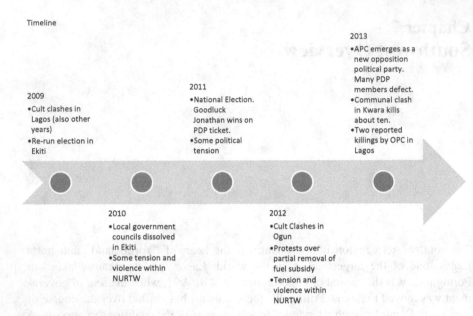

2013
•APC emerges as a
new opposition
political party.
Many PDP
members defect.
•Communal clash
in Kwara kills
about ten.
•Two reported
killings by OPC in
Lagos

2011
•National Election.
Goodluck
Jonathan wins on
PDP ticket.
•Some political
tension

2009
•Cult clashes in
Lagos (also other
years)
•Re-run election in
Ekiti

2010
•Local government
councils dissolved
in Ekiti
•Some tension and
violence within
NURTW

2012
•Cult Clashes in
Ogun
•Protests over
partial removal of
fuel subsidy
•Tension and
violence within
NURTW

Fig. 7.1 Timeline Southwest, Nigeria

rate of per 3.2 % per year. Even though economic conditions of the region are better than the other parts of the country due to high levels of industry and trade, unemployment is high and about two-fifths of the population lives in overcrowded conditions, a quarter without access to basic sanitation (Action on Armed Violence, 2014).

Other states in the region (Ekiti, Kwara, Ogun, Osun, and Oyo) were relatively peaceful during this period, although there was some political violence in 2011 and a communal clash in Kwara in 2013 that reportedly killed about ten people (Fig. 7.1).

7.1 Lagos State

Incidents Per Capita Rank 3/37; Fatalities Per Capita Rank 13/37 (see Fig. 7.2)

One of the six states that makes up the southwestern region of Nigeria, Lagos shares borders with the Republic of Benin to the southwest, as well as a border with the Ogun State. Despite being the smallest state in Nigeria with an area of 3,496 km², the state created in 1967 still has one of the highest populations of the country. Indeed, according to UN estimations, Lagos will become the third largest mega-city in the world, behind Tokyo and Mumbai. Metropolitan Lagos is in fact home to over 85 % of the total population of the state.

As a major metropolis, Lagos is a highly heterogeneous state, comprising of many different ethnic groups. The Yoruba are considered as the main ethnic group, their language being spoken by many of the inhabitants of Lagos. The state is also

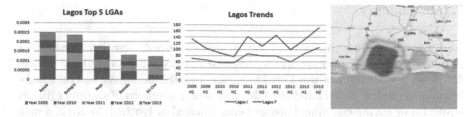

Fig. 7.2 Bar Chart shows annual per capita incidents for Top LGAs; Line Graph shows trend in # of incidents and fatalities in the state; Heat Map shows hotspots of violence from 2009 to 2013—Nigeria Watch data formatted and uploaded to Peace Map (www.p4p-nigerdelta.org/peace-building-map)

home to important international populations, such as the Lebanese, Indian and British communities (www.lagosstate.gov.ng).

Main commercial activities of the state include fishing, both in inland waterways and in the deep-sea, as well as shrimping along the coast. Similarly to many other Nigerian states, agriculture also has a great importance in Lagos' economy, with the main crops being corn, vegetables, rice, yama and coconuts. The metropolis of Lagos is the commercial and financial center of Nigeria.

Bola Tinubu, of the AD, was governor of Lagos State from 1999 to 2007. Babatunde Fashola, also of the AD, was elected in 2007 and reelected in 2011. In 2013 the party joined with other opposition parties to form the APC. The following LGA-level breakdown describes risk factor in the five most violence-prone LGAs in the state. The incidents described are not exhaustive but give a brief overview of the scope and tenor of the types of issues reported in the data during the five year period of 2009–2013.

7.1.1 Apapa

Apapa LGA had the highest number of reported incidents per capita in Lagos State between 2009 and 2013. These incidents included armed robbery, fights between individuals, clashes between robbers and police, and several murders. People were clubbed, shot, sliced, and stabbed with a variety of tools, including guns, bottles, scissors, a razor, and a wheel spanner. In 2010 there was a clash between factions of the Transport Union. Also a fight between two individuals deteriorated into an ethnic clash between Yoruba and Hausa, during which a person was killed. In 2011, two were killed when a disagreement between two individuals escalated into a clash between the AJ1 and Senior Boys gangs. In 2012 road rage escalated into a stabbing, after which the perpetrator was lynched by a mob. In 2013 there was a protest against an oil firm's employment practices and two reported killings by the Oodua People's Congress (OPC), a Yoruba nationalist organization.

7.1.2   Badagry

As in Apapa, the vast majority of the incidents reported were interpersonal or criminal in nature. There were also reports of sea piracy and mob/vigilante justice. In 2009 a suspected pirate was killed by police. In 2012 there were several attacks by pirates reported. In 2012 and 2013 there were several reports of security officials suspected of having committed abuses being attacked or lynched. In 2012, a man was reportedly killed by a policeman for not paying enough of a bribe. In response to the killing, an angry mob reportedly attacked and vandalized the police station. In 2013 a policeman reportedly killed a trader and then was lynched by a mob. Later that same year, a mob reportedly lynched a customs officer after the killing of a motorist. Other incidents in Badagry included the killing of an advisor to the former governor in 2010 and a clash between police and soldiers which killed several in 2011. Also in 2011 there was a clash between factions of a transport union.

7.1.3   Ikeja

In 2009, in addition to reports of criminality, there were many reports of interpersonal violence in which people killed their child, client, roommate, stepbrother, brother, and other acquaintances. In 2010, there was a protest by the Coalition of Youth for Good Governance (CYGG). A Lagos State Traffic Management Authority (LASTMA) official was badly beaten by a mob after he reportedly killed a motorist. In 2013 several members of the Vigilante Group of Nigeria (VGN) were killed by robbers. In addition to reports of murders and criminality, there were several protests in 2013, including one by the workers of the Power Holding Company of Nigeria (PHCN) and another by the Women Arise Initiative. Muslims protested the banning of the hijab in schools. Traders protested the planned demolition of a market. Employees of a manufacturing company protested dangerous working conditions. There was also a clash between the Nigerian Security and Civil Defense Corps (NSCDC) and police. A British citizen was reportedly kidnapped during the year.

7.1.4   Ikorodu

In all years there were multiple reports of murders, criminality, and violence against children. In 2009, 2011, 2012, and 2013 there were clashes reported between two cult groups, the Black Axe and Eiye Confraternities. In 2013 there was also a clash involving the Eiye and KK cult groups. In 2010 there was a reported bus hijacking, allegedly by ritual killers. In 2013 an intra-communal clash was reported in which three people were reportedly killed. After police reportedly killed a motorcycle taxi driver, there was a protest, with was dispersed by security forces.

7.1.5 Eti-Osa

In addition to reports of armed robbery, murder, and domestic violence, there was a chieftaincy tussle in 2009. In 2011 there were reports of a clash that killed two between water vendors and the "area boys" who were extorting them. In 2012 and 2013 there were reports of inter-communal clashes between the Ilaje and Ajah youth during which at least two were killed. Separately, an ocean surge killed about ten people in 2012.

7.2 Kwara State

Incidents Per Capita Rank 17/37; Fatalities Per Capita Rank 24/37 (see Fig. 7.3)

Kwara state borders the Republic of Benin, as well as the states of Niger in the north, Oyo in the southwest, Osun and Ekiti in the southeast and Kogi in the east. Created in 1967, the state is ethnically diverse including the Nupe, Fulani, Bariba and Yoruba (www.kwarastate.gov.ng). Agriculture and animal husbandry are two major commercial activities in Kwara, the main crops grown including rice, corn and soya. Mineral resources include limestone, talc and granite. The state is also well-known for its trademark leather industries and handmade ceramic pottery. With the transition to civilian rule in 1999, Mohammed Lawal, of the ANPP, was elected governor. He was succeeded by Bukola Saraki of the PDP in 2003. Abdulfatah Ahmed, also of the PDP, was elected in 2011. Violence in Kwara has been relatively low over the course of these five years, with an increase in fatalities in 2013, especially in Offa during a communal clash, associated with a longstanding land dispute and a separate bank robbery that reportedly killed about 20. The following LGA-level breakdown describes risk factors in the five most violence-prone LGAs in the state. The incidents described are not exhaustive but give a brief overview of the scope and tenor of the types of issues reported in the data during the five year period of 2009–2013.

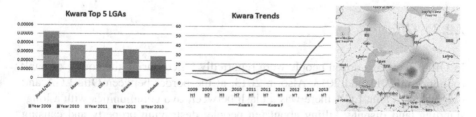

Fig. 7.3 Bar Chart shows annual per capita incidents for top LGAs; Line Graph shows trend in # of incidents and fatalities in the state; Heat Map shows hotspots of violence from 2009 to 2013—Nigeria Watch data formatted and uploaded to Peace Map (www.p4p-nigerdelta.org/peace-building-map)

7.2.1 *Ilorin East/West/South*

These three Local Government Areas surround the city of Ilorin. With almost a million people, Ilorin is among the largest cities in Nigeria, although much smaller than nearby megacity Lagos. As with other densely populated areas, the level of violence tends to be higher than in the surrounding, more rural LGAs in the state. In 2010, reported incidents of violence ranged from armed robbery to gang violence, and one case in which several people died in a stampede at a PDP rally where philanthropists were handing out gifts for the annual Sallah festival. In 2011, an election year, a number of political thugs were reportedly arrested. There was a youth protest over a road which was uncompleted despite the contractor reportedly having been paid for the work. A few people were reportedly killed by suspected ritualists. Several women were murdered during the year. In 2012, a few people were killed in protests against the removal of the fuel subsidy. Other issues in the three LGAs included suspected ritual killings, a political clash, as well as ethnic and cult clashes at the Kwara State Polytechnic University. There was also a student protest during the year over the lack of power supply. In a separate incident, vigilantes clashed with suspected hoodlums, killing several. In 2013, there was a case of a suspected ritual killing of a child, and a student protest over not having received their results in time to participate in the National Youth Service (NYS). Later in the year, a student was reportedly shot by a police officer, which led to more protests. There were a couple of cult clashes, including one between members of the security forces and the Eiye Confraternity, leading to several deaths. In August, a suspected kidnapper was reportedly lynched by a mob.

7.2.2 *Moro*

In the small LGA of Moro, there were a couple of murders reported during the five years period.

7.2.3 *Offa*

In Offa, about two dozen people were reportedly killed during the course of bank robberies in 2011 and 2013. Also in 2013, at the border between Offa LGA and neighboring Oyun LGA there was a clash between two communities with a longstanding land dispute, killing about ten people, destroying property and causing people to flee. The incident was triggered by a car accident, which led to an argument and escalated from there.

7.2.4   *Kaiama*

Several murders, including a suspected ritual murder of a child, were reported in Kaiama during the five year period.

7.2.5   *Ifelodun*

The majority of the incidents reported in Ifeludun LGA between 2009 and 2013 were related to communal conflict issues. In 2010 a longstanding chieftaincy tussle between two clans reportedly escalated over a disagreement concerning who would lead the Eid prayers. Then, in 2013 conflict escalated again in the same community killing at least three.

7.3   Oyo State

Incidents Per Capita Rank 26/37; Fatalities Per Capita Rank 30/37 (see Fig. 7.4)

Oyo State borders neighboring Benin to the west. Within the country of Nigeria, it borders the states of Kwara, Osun, and Ogun. It is less ethnically diverse than many other states, with the majority of the population being of Yoruba descent. The economy is predominantly agricultural. On a per capita basis, it has registered among the lowest levels of violence in the country (www.oyostate.gov.ng).

Unlike most states, the governorship of Oyo has switched parties several times since the beginning the Fourth republic. In 1999 Alhaji Lamidi Adesina was elected under on the AD (Alliance for Democracy) platform. Then in 2003, Rashidi Adewolu Ladoja, from the PDP, was elected. He was impeached for corruption in 2006, then reinstated briefly after the impeachment was overturned. In 2007, Christopher Alao-Akala, also of the PDP, was elected. In 2011, Abiola Adeyemi Ajimobi won the governorship as a member of the ACN party, which is regionally powerful in the Southwestern part of the country, including Lagos. The following LGA-level breakdown describes risk factors in the five most violence-prone LGAs in the state. The incidents described are not exhaustive but give a brief overview of the scope and tenor of the types of issues reported in the data during the five year period of 2009–2013.

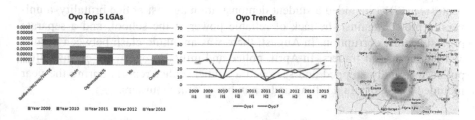

Fig. 7.4 Bar Chart shows annual per capita incidents for top LGAs; Line Graph shows trend in # of incidents and fatalities in the state; Heat Map shows hotspots of violence from 2009 to 2013—Nigeria Watch data formatted and uploaded to Peace Map (www.p4p-nigerdelta.org/peace-building-map)

7.3.1   Ibadan North/Northeast/Northwest/Southwest/Southeast

The capital city of Oyo State, Ibadan, is the fourth largest city in the country, after Lagos, Kano, and Abuja. The greater Ibadan area is made up of five densely populated LGAs. In Oyo State the vast majority of all conflict incidents were reported in this area. In 2009, most incidents were criminal. Among those, there were a couple cases of violence against women and girls, and some suspected ritual killings. Additionally, there was one student protest and a case of political violence between supporters of the PDP and the ACN. In 2010, beyond the reports of criminal and interpersonal violence, there were reports of tension and violence within the NURTW (National Union of Road Transport Workers). There was a student protest over religious issues which temporarily halted classes. A politician was reportedly killed during the year. In 2011, an election year, violence took on a more political character as violence within, or involving, the NURTW escalated. There was a reported clash between supporters of the PDP and the AP (Accord Party) in April. In July there was reportedly a university protest over the lack of water and electricity. In 2012, in addition to murders and other reports of criminal violence, there were a number of protests and political flare-ups. Protests included one against the partial removal of the fuel subsidy, a youth protest over the alleged killing of a man by a police officer, and a women's demonstration in support of legislation against violence against women and girls. A clash reportedly broke out between motorcycle taxi (okada) drivers and police, after a driver was arrested for carrying more than one passenger. Several people were injured in the incident. NURTW members were involved in violence during the year as well, the most severe of which was a clash between two NURTW factions that killed two. In October, it was reported that NURTW members in support of the ACN governor had a fight with PDP supporters, causing minor injuries. Other incidents included a ritual killing, and rough handling of a journalist by security forces during a political event. In 2013, there were incidents of political and ethnic violence as well as a few student protests. NURTW was involved in violence as in previous years. The deputy chairman of the NURTW was reportedly shot dead. In another incident NURTW members reportedly clashed with ACN protesters seeking the removal of the state party chairman. NURTW members, together with ACN supporters, reportedly clashed with AP members, injuring several. A NURTW branch chairman was murdered at a masquerade festival in Ibadan. Gunmen reportedly attempted to kill an AP chieftain. His night guard was killed in the incident. Protests in 2013 included a student demonstration against police brutality, a university protest against the lack of reliable power at the school, and a separate protest against corruption. In August, there was an ethnic clash between Yoruba and Hausa market traders. In April, police and military reportedly clashed after police reportedly stopped and searched a soldier. Other reports during the year included a case of ritual killing and a number of other murders.

7.3.2 Iseyin

During the five year period, there were a few incidents of criminal violence (car smugglers clashing with customs officers, armed highway robbery, and bank robberies. A PDP leader was killed in 2010. A suspected ritualist was lynched in 2012.

7.3.3 Ogbomosho North/South

Most incidents in these two LGAs related to criminal violence by armed robbers including a series of bank robberies that led to the death of seven. Additionally, however, there was a case of a political rally that turned violent, the lynching of a ballot box snatcher, a kidnapping, and the killing of a student by a university cult group.

7.3.4 Ido

In Ido LGA, the issue reported primarily dealt with domestic and interpersonal violence.

7.3.5 Orelope

In 2009 there was reported to be an inter-communal clash over the ownership of a market, which killed one person. In 2011 there was a reported political clash between PDP and ACN which killed two.

7.4 Ogun State

Incidents Per Capita Rank 7/37; Fatalities Per Capita Rank 14/37 (see Fig. 7.5)

Ogun borders the Republic of Benin in the west. Within Nigeria, it borders Lagos to the south, Ondo and Osun to the east and Oyo to the north. Demographically, the population is predominately Yoruba, composed mainly of the Egba, Yew, Awori, Egun, Ijebu and Remo clans.

The main source of income for the state of Ogun is agriculture, mostly the cultivation and growth of cocoa, kola nut, rice, maize, banana, and palm oil. Mineral industries include limestone, phosphate, glass, clay, and granite (www.ogunstate. gov.ng). Ogun also enjoys a number of touristic attractions, monuments, museums and festivals, such as the Olumo Rock, which is one of the most popular tourist destinations in Nigeria and West Africa, and the Ebute Oni Beach. Among the several festivals taking place in Ogun state, the two most famous ones are Ojude Oba, a carnival-like celebration, and the Agemo celebrated in June/July every year.

Fig. 7.5 Bar Chart shows annual per capita incidents for top LGAs; Line Graph shows trend in # of incidents and fatalities in the state; Heat Map shows hotspots of violence from 2009 to 2013—Nigeria Watch data formatted and uploaded to Peace Map (www.p4p-nigerdelta.org/peace-building-map)

As with other states in the region the governorship of the state of Ogun has shifted from AD to PDP to ACN over the course of the Fourth Republic. The first governor, Olusegun Osoba (AD), was elected in 1999. In 2003, the PDP won the governorship with the election of Gbenga Daniel, who served until 2011. He was succeeded by Ibikunle Amosun (ACN).

Violence in Ogun primarily had to do with armed robbery and interpersonal violence. But there were also significant levels of cult violence, clashes between factions of trade unions, some political violence, student protests, attacks on energy infrastructure, and other issues. The following LGA-level breakdown describes risk factors in the five most violence-prone LGAs in the state. The incidents described are not exhaustive but give a brief overview of the scope and tenor of the types of issues reported in the data during the five year period of 2009–2013.

7.4.1 Ijebu Ode

In 2009, issues of armed robbery and interpersonal violence were reported. In 2010, one person was killed in cult violence between the Eiye cult group and the Aye cult group. Cultists also reportedly killed a lecturer. Most incidents in 2011 had to do with cultism in Ogun universities. In July, two women were reportedly raped by members of the Eiye cult group at the Tai Solarin University of Education. In the protests which followed the incident, a student was reportedly shot by police. In October, Eiye, Aiye, and Lord Cult groups reportedly clashed, leading to the deaths of at least five people. In November a clash between NURTW factions reportedly killed three. In 2012, cult violence between the Eiye and the Aiye groups continued, killing three people. In a separate incident, the wife of a lawmaker was reportedly kidnapped. In 2013, a student protest over tuition fees and examinations turned violent with multiple arrests and allegations of police brutality. Two months later, suspected cultists killed at least two people.

7.4.2   Remo North

In Remo North, reported incidents in the five year period had to do with murder and armed robbery.

7.4.3   Abeokuta North/South

Abeokuta, the capital city of Ogun, and the 15th largest city in the country, comprises two LGAs. As with many urban centers, most reported incidents are criminal, but some have a collective violence aspect to them as well. In 2009, in addition to multiple cases of fatalities associated with armed robbery, there was also a case of NURTW violence at a motor park. Separately, two suspected ritualists (human parts dealers) were lynched by a mob. In 2010, an NURTW secretary was murdered. There were a couple of protests, including one by Postal Service retires over compensation. In 2011, a clash between factions of the ACO-MORAN (Amalgamated Commercial Motorcycle Riders Association of Nigeria) reportedly killed several people. A member of the ACN was reportedly killed for political reasons. In 2012, a protest over the partial removal of the fuel subsidy reportedly turned violent. Cult violence involving Eiye at MAPOLY (Mashood Abiola Polytechnic), reportedly killed one person. In 2013, masqueraders reportedly killed an Imam. TOAN (Truck Owners Association of Nigeria) reportedly staged a protest for fewer restrictions on their vehicles' movement. Two factions of the RTEAN (Road Transport Employers Association of Nigeria) clashed, injuring several people. Vigilantes reportedly killed two suspected robbers. Women and children protested underage marriage laws. Rival factions of the APC (All Progressive Congress) clashed, injuring dozens. Later in the year, an APC leader was attacked in his home. Women at MAPOLY protested against the rape of a student. Commercial motorcycle (Okada) riders, with a legal dispute, clashed, injuring several.

7.4.4   Obafemi Owode

Two were killed in 2009 when the Navy fought with local youths. Later in the year a NNPC (Nigeria National Petroleum Corporation) pipeline was reportedly vandalized, leading to an explosion. In 2010, two PDP factions reportedly clashed, killing several. In 2012 at least one person was reportedly killed during a protest against the partial removal of the fuel subsidy. Oil thieves reportedly killed three NNPC workers. In 2013, a violent clash between police and pipeline vandals killed several.

7.4.5   Ewekoro

Incidents reported in Ewekoro related mostly to armed robbery and other forms of criminal violence.

7.5   Osun State

Incidents Per Capita Rank 29/37; Fatalities Per Capita Rank 33/37 (see Fig. 7.6)

Osun is a small, predominantly Yoruba state, state bordering Ondo State to the southeast, Kware to the north, Ekiti to the northeast, Oyo to the west and Ogun to the southwest. The primary source of income of Osun is agriculture. Agricultural products include food crops such as yam, cassava, maize, vegetables, fruits and beans, and cash crops including cocoa and kola nuts. Minerals include gold, clay, limestone, and granite. Osun is well-known for its traditional fabric and batik. Tourist attractions include the UNESCO World Heritage Cultural site of the Osun-Osogbo Sacred grove which contains shrines and sculptures in honor of Osun, the Yoruba goddess of fertility, as well as other deities. Other examples include the ancient city of Ile Ife, the Oranmiyan Staff (a 5 m-high staff which belonged to the third king of Ife), the Ibodi Monkey Forest, and the Oke Maria ("Mary's Mountain) which is visited every year by Catholics as a pilgrimage in February (www.osun.gov.ng).

Like some of the other states in the Southwest, the politics of the state have evolved since the Fourth Republic began in 1999. The first civilian governor was Adebisi Akande, elected on the AD platform. He was succeeded in 2003 by Olagunsoye Oyinlola of the PDP. Osun's governorship has been held by a member of the PDP. Oyinlola initially was declared to have won reelection in 2007, but the results were overturned by the courts in 2010, and Rauf Aregbesola of the ACN was declared the winner.

Osun is one of the least violent states in Nigeria, according to the incident data collected. The following LGA-level breakdown describes risk factors in the five most violence-prone LGAs in the state. The incidents described are not exhaustive but give a brief overview of the scope and tenor of the types of issues reported in the data during the five year period of 2009–2013.

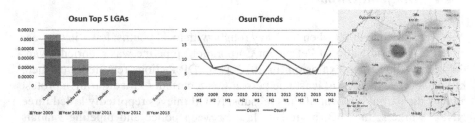

Fig. 7.6 Bar Chart shows annual per capita incidents for top LGAs; Line Graph shows trend in # of incidents and fatalities in the state; Heat Map shows hotspots of violence from 2009 to 2013—Nigeria Watch data formatted and uploaded to Peace Map (www.p4p-nigerdelta.org/peace-building-map)

7.5.1   Osogbo

In 2009, most incidents were domestic and criminal murders as well as a few incidents in which police officers allegedly killed people when they wouldn't pay a
bribe. In 2010, an intoxicated police officer allegedly killed two men and was in turn
killed by an angry mob. In 2012 there were a number of protests, including one over
the partial removal of the fuel subsidy. A clash between two cult groups was also
reported in 2012. Most incidents in 2013 related to protests, including a number
related to education policy and labor strikes. Additionally, PDP members reportedly
protested the demolition of a party building, and civil servants protested over their
pension payments.

7.5.2   Ilesha East/West

Armed robbery and murder were the main issues reported in Ilesha East and West.
Two robbers were reportedly lynched by a mob in June. Also in 2009, two political
factions clashed, killing three. Political violence continued into 2010 when dozens
of thugs stormed a political meeting. In 2012, okada (commercial motorcycle) drivers rioted when a driver was alleged to have been killed by police. In a separate
incident, a journalist was reported to have been killed.

7.5.3   Obokun

In Obokun, incidents were mainly interpersonal and criminal. In 2013, however,
there were unconfirmed reports of a gang attacking the palace of a traditional ruler,
killing him, and burning it down.

7.5.4   Ila

In 2012, youths reportedly protested the deaths of their colleagues in a road accident
by burning the LGA secretariat.

7.5.5   Ifelodun

In 2011 a former NLC (National Labour Congress) chairman was killed. In 2012
there was a clash reported between two cult groups.

7.6   Ekiti State

Incidents Per Capita Rank 22/37; Fatalities Per Capita Rank 31/37 (see Fig. 7.7)

Created in 1996 out of Ondo state, Ekiti state is the sixth smallest and eighth least populous (not including the FCT) state in Nigeria. It is located in the southwestern part of the country, and borders Kwara to the north, Osun to the west, Ondo to the south, and Kogi to the east. Its approximately 2.4 million people (2006 census) are predominately of Yoruba descent. Agriculture is the backbone of the state's economy, producing both cash and food crops, with the most significant being cocoa, palm oil, kola nuts, plantains, bananas, cashews, citrus, rice, yam, cassava, maize, and cowpeas. Mineral sources include mineral resources, including bauxite and tin ore (www.ekitistate.gov.ng).

Between the years 2009 and 2013, Ekiti was one of the least violent states in Nigeria, with a peak in the first half of 2011. The violence in the state was a mix of crime and violence between political groups, especially in the run up to the 2011 elections, which took place in the first half of that year, after which violence in general and clashes between political groups specifically dropped dramatically, with the remaining violence consisting mostly of just petty crime.

Like several other states in the region, the first civilian governor was elected in 1999 on the AD platform. Then in 2003, Peter Ayodele Fayose from the PDP was elected but later impeached and replaced by Tunji Olurin, also of the PDP. In 2007, Olusegun Oni (PDP) was elected, though the results were invalidated leading to a rerun in 2009. In 2010 Kayode Fayemi assumed office on the ACN platform. The following LGA-level breakdown describes risk factors in the five most violence-prone LGAs in the state. The incidents described are not exhaustive but give a brief overview of the scope and tenor of the issues reported in the data during the five year period of 2009–2013.

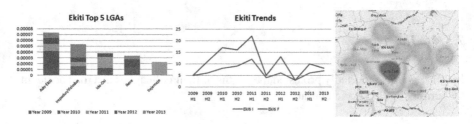

Fig. 7.7 Bar Chart shows annual per capita incidents for top LGAs; Line Graph shows trend in # of incidents and fatalities in the state; Heat Map shows hotspots of violence from 2009 to 2013—Nigeria Watch data formatted and uploaded to Peace Map (www.p4p-nigerdelta.org/peace-building-map)

7.6.1   Ado-Ekiti

In 2009, there were violent protests after the murder of a political aspirant. After the 2007 election was invalidated, political tensions were high in run-up to the 2009 re-run. Protests were reported when the release of the results of the rerun were initially delayed. Also in 2009 there was a clash reported between two transport unions. Separately, a gang reportedly killed a federal SARS (Special Anti-Robbery Squad) official. In 2010, most reported issues related to cult violence, including clashes between rival gangs. Violence by cultists reportedly included torture, rape, murder, and kidnapping. Students protested the increase in school fees at the University of Ado Ekiti. Anti-union activists reportedly killed a PDP supporter. Governor Fayemi dissolved the state's local government councils. In 2011, most reports had to do with fatalities associated with armed robbery, but later in the year there was also a suspected bomb explosion at the INEC (Independent National Electoral Commission) office. In 2012 there were several protests, including a peaceful protest by market women supporting the recommendation of the NJC (National Justice Commission), and a protest by the NYSC (National Youth Service Corps) over the payment of their allowance. In September, university students reportedly rioted when a youth was hit by a car. There was a clash reported between supporters of the PDP and supporters of the ACN. In 2013, youths protested the opening of a market. NANS (National Association of Nigerian Students) protested in support of the ASUU (Academic Staff Union of Universities) labor strike demands. Transport union members clashed with police over the payment of a fare. Students protested over a university policy matter. In a separate incident, journalists and PDP leaders were attacked by gunmen. PDP youths protested against the party's choice of candidate in an upcoming election.

7.6.2   Irepodun/Ifelodun

In addition to a number of incidents of robbery, rape and murder, there was a protest by local government workers for implementation of the minimum wage law in 2012. In 2013 there was a politically charged clash between two factions at a funeral.

7.6.3   Ido-Osi

A number of murders were reported in Ido-Osi during the five year period.

7.6.4   Ikere

Rapes, robberies, abductions, and a few protests over electricity bills and working conditions were reported during the five year period.

7.6.5   Ilejemeje

An ACN member was reportedly shot dead in the election year of 2011.

Chapter 8
South Central Overview

The level of violence was relatively low in the South Central region of Nigeria between the years of 2009 and 2013. On average, according to Nigeria Watch (data formatted on the P4P Peace Map), there was an increase in political violence during the 2011 election year, but no state ranked higher than 16 in reported fatalities per capita over the five year period.

With the exception of Kogi State, most people in this region are ethnically Igbo. The region's population is predominantly Christian.

Historically, Enugu was the capital of the secessionist Biafran state, which attempted to break away during Biafran War of 1967–1970, killing over a million people. During this book's period of interest (2009–2013), the legacy of that war continued to reverberate in the region with agitations by the Movement for the Actualization of the Sovereign State of Biafra (MASSOB), especially in Anambra and Ebonyi, as well as neighboring Imo State in the Niger Delta region. MASSOB was created in 1999 by Chief Ralph Uwazurike, as an attempt to non-violently resurrect the struggle for self-determination waged by the Igbo of south east Nigeria (Falola and Matthew 2008). The organization also desired to ensure integration of the Igbo minority into the post-war Nigerian society. Currently, individuals involved in violence who are associated with MASSOB in press reports are often unemployed, uneducated, dropout youths committing crimes both as individuals and in gangs and "cults". Between 2009 and 2013 incidents involving MASSOB usually involved violent protests against alleged police corruption or abuses. In 2013 violence broke out between MASSOB and the Association of Igbo Youths Organization (AIYO) after a period of jostling over which group would be the vanguard of the cause.

Another way in which the Biafran War still resonates is with occasional spikes of ethnic tension between the Igbo and the Hausa, such as in 2012 in Anambra after a police officer allegedly killed a bus driver for not paying a bribe. Other unrelated issues in the region included communal violence, criminality, cult violence involving the Black Axe and Blue Beret confraternities, and political tensions during the election year of 2011. Kogi state, which is socio-politically quite different from the other

© Springer International Publishing Switzerland 2015
P. Taft, N. Haken, *Violence in Nigeria*, Terrorism, Security,
and Computation, DOI 10.1007/978-3-319-14935-6_8

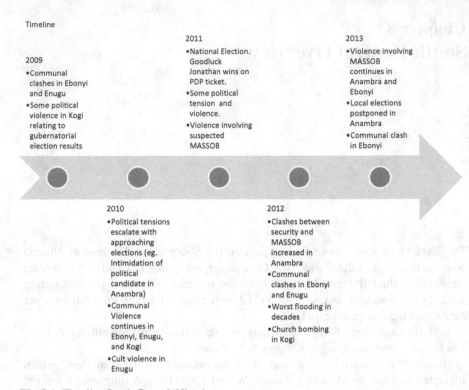

Timeline

2009
• Communal clashes in Ebonyi and Enugu
• Some political violence in Kogi relating to gubernatorial election results

2011
• National Election. Goodluck Jonathan wins on PDP ticket.
• Some political tension and violence.
• Violence involving suspected MASSOB

2013
• Violence involving MASSOB continues in Anambra and Ebonyi
• Local elections postponed in Anambra
• Communal clash in Ebonyi

2010
• Political tensions escalate with approaching elections (eg. Intimidation of political candidate in Anambra)
• Communal Violence continues in Ebonyi, Enugu, and Kogi
• Cult violence in Enugu

2012
• Clashes between security and MASSOB increased in Anambra
• Communal clashes in Ebonyi and Enugu
• Worst flooding in decades
• Church bombing in Kogi

Fig. 8.1 Timeline South Central, Nigeria

states in this region, had the fewest incidents and fatalities reported per capita. However, there were a few cases of violence reported involving suspected Islamist insurgents, including an attack on a church in Okene in 2012 that killed over a dozen people and several clashes with security forces (Fig. 8.1).

8.1 Anambra State

Incidents Per Capita Rank 13/37; Fatalities Per Capita Rank 16/37 (see Fig. 8.2)

Anambra is a small, densely populated state, bordering Kogi to the north, Edo and Delta to the west, Rivers, and Imo to the south, and Enugu to the east. It has approximately 4.2 million people (2006 census). More than half of Anambra's population (52.9 %) is under the age of 18. Ethnically, the population is predominately of Igbo descent (~98 % of the population), with a small minority of Igala people in the north-western part of the state (www.anambrastate.gov.ng).

Anambra has rich reserves of a variety of natural resources, including natural gas, crude oil, bauxite, and ceramics, supplemented by agriculture, fishing, and animal husbandry. The discovery of oil and natural gas has created a new sector in the economy.

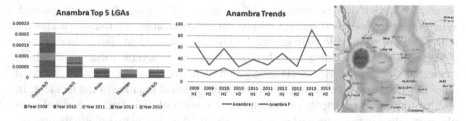

Fig. 8.2 Bar Chart shows annual per capita incidents for top LGAs; Line Graph shows trend in # of incidents and fatalities in the state; Heat Map shows hotspots of violence from 2009 to 2013—Nigeria Watch data formatted and uploaded to Peace Map (www.p4p-nigerdelta.org/peace-building-map)

Despite new opportunities that the discovery of oil poses, the gap between rich and poor has deepened.

Anambra was deeply affected by the 1967–1970 Nigeria–Biafra war. However, during the period of 2009–2013, Anambra had moderate levels of violence, with a spike in fatalities in the first half of 2013 when between 30 and 45 corpses were found floating in a river, with no additional information. Issues reported were primarily of a criminal nature, including armed robbery, murder, and kidnapping.

In addition to incidents of criminality there were also a few inter-communal clashes during the 2009–2013 period, most frequently related to land disputes. There were a few reported clashes between the security forces and the Movement for the Actualization of the Sovereign State of Biafra (MASSOB).

Politics in Anambra have been quite dynamic since the early days of the Fourth Republic. In 1999, Chinwoke Mbadinuju from the ruling PDP was elected as governor. Then in 2003, the PDP won again, but after years of litigation, the election was overturned and Peter Obi of the All Progressives Grand Alliance (APGA) took the governorship in 2006. Obi was subsequently impeached briefly, but the impeachment was overturned, and he remained in office until 2014. The following LGA-level breakdown describes risk factors in the five most violence-prone LGAs in the state. The incidents described are not exhaustive but give a brief overview of the scope and tenor of the types of issues reported in the data during the five year period of 2009–2013.

8.1.1 Onitsha North/West

In 2009 several people were killed in the course of robberies by gangs of gunmen. There were at least three reports of mob lynchings of people accused of stealing items such as purses, phones, and money. Some people were also killed in the course of kidnappings including the abduction of traditional rulers, businessmen, and a child. In 2010, there were more reports of lynchings, including that of a madman who had attacked a woman, and another of a suspected thief. Twice, traffic police were accused of killing drivers when they believed their directives were being ignored. When fees were levied on motorcycle taxis in a market, angry motorcyclists clashed with police. Three people were reportedly killed in the incident. Two police officers were killed during a kidnapping. In 2011, there were several incidents in which suspected members of MASSOB were killed by police. In some cases the deceased were reported to be committing robberies or extortion at the time of the incident. In another they were reportedly protesting the detention of their leaders. Several people were killed in the course of robberies. About 20 bodies were reportedly discovered in a tunnel, killed by a gang of armed robbers. In 2012, clashes between security forces and MASSOB escalated, with incidents reported in January, February, June, and July. In the February incident it was reported that a police officer killed a bus driver for failing to pay the normal "tip." Because the police officer was said to be Hausa, it led to ethnic unrest. MASSOB also mobilized around the incident, but focused their energy on attacking police checkpoints. In the June incident, security forces raided the MASSOB office. In the subsequent clash, over a dozen people were reportedly killed. Other incidents during the year included fatalities associated with robberies and several kidnappings. In the floods of 2012, about ten people were reportedly drowned. In 2013 clashes between MASSOB and security forces continued. In April and July, there were also clashes reported between MASSOB and the Association of Ibgo Youths Organization (AIYO). A suspected robber was reportedly lynched by a mob in August. Separately, a member of the Anambra State Traffic Agency (ASTA) was reportedly killed by a vigilante. A couple of people were reportedly killed when youths attempted to impose levies on motorists at a motor-park.

8.1.2   Awka North/South

In 2009, most incidents related to criminality (robberies, kidnapping, and murders). In 2010, the PDP candidate for governor, Chukwuma Soludo was threatened by angry youth supporters of APGA. Police intervened and rescued him from a possible lynching. In 2011 APGA supporters clashed with supporters of the AP (Accord Party) at a rally for AP senator Annie Okonkwo. In 2012, protests by students and civil servants were reported, as well as several incidents of criminality. One of the most widely reported incidents of 2013 was in January when between 30 and 45 bodies were found floating down the Ezu River. The identities of the perpetrators were never established but MASSOB accused the security services of extra-judicial killing. In June there was a clash reported between police and MASSOB. Separately, several people were reportedly kidnapped during the year, including two politicians, a monarch, and the wife of the Anambra state Labour Party chairman. Market traders protested a tax in May. In November, protesters demonstrated against the local election. In one of these protests two women were reportedly killed when police intervened. Separately, university staff protested inadequate funding. In July there was a clash between community youths and a Catholic church over a disagreement over burial traditions. Later, in August there was an unrelated clash between traditional masqueraders and Christian worshipers that led to the death of one person and the burning of several churches. In September the Independent National Electoral Commission postponed local elections in Anambra, leading to protests by candidates at Government House Awka. Multiple murders were also reported during the year.

8.1.3 Ihiala

From 2009 to 2011 the issues reported in Ihiala were mainly murders and bank robberies. In 2012, there was an increase in reports of kidnappings (including that of the deputy chairman of Ihiala LGA) and clashes between security forces and kidnappers. In 2013 there were a couple reports of murders and a clash between an armed gang and the Special Anti-Robbery Squad (SARS).

8.1.4 Ekwusigo

In 2009 a pastor was reportedly kidnapped and killed. In 2010, police reportedly killed a suspected kidnapper of the former Commissioner for Health. In 2011, a political candidate was reportedly abducted. In 2012 police clashed with gunmen and kidnapping suspects, killing several. In 2013 there was a report of a murder in Ekwusigo.

8.1.5 Idemili North/South

In Idemili North and Idemili South there were a few incidents in 2009 associated with criminality and vigilante violence. A movie star was reportedly abducted and a police officer killed. In 2010 a few murders were reported. Also there was an intra-communal clash over land and a political clash when thugs attacked the PDP gubernatorial candidate's convoy. In 2011, several reported incidents related to clashes between security forces and suspected criminals. Hundreds of community youths reportedly protested violently in an attempt to ensure that their names were listed as members of the community. In June there was reportedly a clash between market traders and the security forces that injured dozens. An explosion at a monarch's palace was suspected to be caused by a bomb. In 2012 there was a clash between union factions over marketplace levies. Several were killed in a clash between vigilantes and gunmen. A case of inter-communal violence over land was reported. Several people were reportedly killed during a cultural festival. In 2013, several murders were reported, including that of a community chairman and a businessman. A traditional ruler survived an attempted murder. Clashes between MASSOB and police reportedly injured about ten and killed about four. Other incidents reported during the year included the abduction of a monarch, the ritual killing of a woman, and the rape and murder of a woman.

8.2   Ebonyi State

Incidents Per Capita Rank 15/37; Fatalities Per Capita Rank 17/37 (see Fig. 8.3)

Established in 1996 from parts of Enugu and Abia states, Ebonyi state is the fourth smallest and third least populous (excluding the FCT) state in Nigeria. It is located in southeastern Nigeria, bordering Benue to the north, Enugu to the west, Abia to the south, and Cross River to the east. Its population is relatively ethno-linguistically homogenous, comprised primarily of Igbo. The state's economy is primarily agricultural, and is a leading producer of rice, yam, potatoes, maize, beans, and cassava. Mineral resources include lead, crude oil, and natural gas (www.ebonyionline.com/about-ebonyi-state).

Ebonyi was a moderately violent state between 2009 and 2013, with land conflicts and inter-communal violence, relating to clan differences, socio-political inequality and agrarian economic interests. Ebonyi state was marked by repeated clashes between the Ezillo and Ezza communities in Ishielu, principally over land disputes. In one incident, between 50 and 60 people were killed, mostly women, children, and the elderly. Markets and houses were burned. Livestock was slaughtered.

Another contributor to violence is the role of university-based confraternity gangs known as "cults" that are often reported at Ebonyi State University in Abakaliki. Renewed violence in 2012 and 2013 between University cults resulted in gun battles, kidnappings, and murder, including the killing of a police corporal in 2012 and multiple deaths in 2013.

The state had a PDP governor since the beginning of the Fourth Republic in 1999. In 2007, Martin Alechi was elected governor. He was re-elected in 2011. The following LGA-level breakdown describes risk factors in the five most violence-prone LGAs in the state. The incidents described are not exhaustive but give a brief overview of the scope and tenor of the types of issues reported in the data during the five year period of 2009–2013.

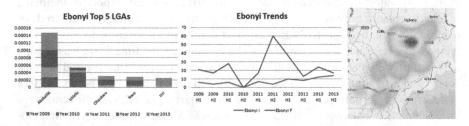

Fig. 8.3 Bar Chart shows annual per capita incidents for top LGAs; Line Graph shows trend in # of incidents and fatalities in the state; Heat Map shows hotspots of violence from 2009 to 2013—Nigeria Watch data formatted and uploaded to Peace Map (www.p4p-nigerdelta.org/peace-building-map)

8.2.1 Abakaliki

In 2009 issues reported mainly had to do with clashes with armed robbers. In 2010, in addition to criminal incidents, there was a political clash between PDP factions. Separately, thousands of Anglicans reportedly protested the administration of the governor in regards to development and poverty. In 2012 there were a number of people reportedly killed in clashes with armed robbers, at least two kidnappings, including that of a government official and of a businessman, and one case of ritual killing. There was also an attack by cultists that reportedly killed three students. In 2013, clashes between cult groups continued, reportedly killing one person. There were also two kidnappings reported, a few murders, and university protests during the year.

8.2.2 Ishielu

In 2009, the conflict between Ezzilo and Ezza reportedly killed several people. The violence continued into 2010, when a reported clash between the two communities killed five. In 2011 the conflict escalated with an incident that reportedly killed dozens. Other incidents between 2009 and 2013 included an intra-communal chieftaincy tussle and several cases of criminal violence.

8.2.3 Ohaukwu

There were no incidents reported in 2009–2010. But in 2011 a mob lynched four suspected rapists and robbers. In 2012, an inter-communal clash with a community in the neighboring state of Benue reportedly killed about ten people. In 2013 police dismantled a gang of kidnappers. There was also one case reported in which members of the Ejilewe Ukwuagba Neighbourhood Security Committee, a vigilante group, tortured a man to death for allegedly stealing a woman's phone.

8.2.4 Ikwo

In 2011 a clash was reported between supporters of the PDP and supporters of ANPP, killing one person. Also reported during the year was a riot triggered by the mysterious death of a pastor. In 2012, there were reports of inter-communal clashes over the ownership of a fishing pond that killed several people. Also during the year, there was a reported ritual killing. Adding to the pressure of the administration of the LGA was severe flooding in the country's worst rainy season in decades. In 2013 an inter-communal clash between a community in Ikwo and one across the border in Cross River reportedly killed several people.

8.2.5 Izzi

Izzi was the fifth most violent LGA in Ebonyi during the five year period. In 2011 there was a political clash between supporters of the PDP and the ANPP. Three people were reportedly killed in the incident. Flooding reportedly killed three in 2012. In 2013, in addition to a number of murders, there was a clash in a market square involving MASSOB and the police.

8.3 Enugu State

Incidents Per Capita Rank 19/37; Fatalities Per Capita Rank 29/37 (see Fig. 8.4)

Located in southeastern Nigeria and former capital of the short-lived Federal Republic of Biafra (1967–1970), Enugu state is the 29th largest and 22nd most populous state in the country, bordering Kogi and Benue to the north, Anambra to the west, Imo and Abia to the south, and Ebonyi to the east. Its approximately 3.3 million people (2006 census) are predominately of Igbo descent.

The economy is primarily rural and agrarian, with yam tubers, palm produce and rice as its main crops. However, a significant part of the state's population is also engaged in the trading and services sectors. Traditional industries include wood carving, blacksmithing pottery, basket and mat making, and cloth weaving and dyeing based on the local cotton production. Mineral resources include coal, limestone, iron ore, crude oil, natural gas, and bauxite (www.enugustate.gov.ng).

Over the five years between 2009 and 2013, Enugu was one of the less violent states in the country in terms of fatalities per capita, although it had a relatively high number of minor incidents. Most violence had to do with petty crime, with a few cases of inter-communal clashes.

Since the Fourth Republic began in 2009, the governor of Enugu has been a member of the PDP. In 2011, Governor Sullivan Chime was reelected to a second term. The following LGA-level breakdown describes risk factors in the five most

Fig. 8.4 Bar Chart shows annual per capita incidents for top LGAs; Line Graph shows trend in # of incidents and fatalities in the state; Heat Map shows hotspots of violence from 2009 to 2013—Nigeria Watch data formatted and uploaded to Peace Map (www.p4p-nigerdelta.org/peace-building-map)

violence-prone LGAs in the state. The incidents described are not exhaustive but give a brief overview of the scope and tenor of the types of issues reported in the data during the five year period of 2009–2013.

8.3.1 Enugu East/North/South

In 2009 two protests reportedly turned violent. Two kidnappings resulting in fatalities were reported. One person died during a mass prison break. A few murders were also reported during the year. In 2010, most incidents reported had to do with violent crime. There were a couple of university protests during the year, in which three people were reportedly killed. There were also two clashes reported between the Black Axe and Blue Beret cult groups, which killed about four. In 2011, a traditional ruler was kidnapped and the chairman of the PDP was shot. In 2012, cult violence reportedly escalated. There were also multiple kidnappings reported, and a protest by Biafran separatists at which many were detained. In 2013, a police commissioner from Kwara State was reportedly killed in Enugu. Other incidents included the killing of a security official by students during a university election and several other murders during the year. There was an armed clash in a market place, reportedly involving a gang of robbers, police, and vigilantes that killed several people. The Igbo Progressives Union (IPU) staged a peaceful protest against the Aviation Minister.

8.3.2 Nsukka

In Nsukka, incidents ranged from riots and protests (2010), kidnapping (2012), cult violence (2011), and several disputes and murders, including the killing of a human rights activist in 2013.

8.3.3 Nkanu East/West

Communal violence was more common in Nkanu East and West than in most of the other Enugu LGAs during the five year period. In 2009, 2012, and 2013 there were clashes reported between Okuru and Umuode communities, which killed a total of about a dozen. In 2010, a man reportedly died in a chieftaincy tussle. In 2012 there was a clash between Fulani pastoralists and a farming community in which a man was killed and several women raped. A paramount ruler was reportedly murdered in 2012.

8.3.4   Igbo-Eze North/South

In 2011, there was reportedly a clash between two communities in the context of traditional Akatakpa masquerades which killed two. Other incidents in the five year period included the killing of a former LGA chairman in 2012.

8.3.5   Udi

As in Nkanu East and West LGAs, Udi experienced a significant amount of communal violence during the five year period. In 2009 an attack on Amachalla Ngwo community was reported, resulting in the death of three. There was also a land conflict reported between Affa and Umulumgbe communities in 2009. In 2010 clashes were reported between Umuoka and Affa communities. In 2012 a herdsman was reportedly found dead after a dispute over grazing land with local farmers. Separately, also in 2012, a police superintendent was reportedly killed.

8.4   Kogi State

Incidents Per Capita Rank 25/37; Fatalities Per Capita Rank 22/37 (see Fig. 8.5)

Formed in 1991 from parts of Kwara state and Benue state, Kogi is the 13th largest and the 20th most populous state in Nigeria. It borders Nasarawa, Niger, and the FCT to the north, Kwara and Ekiti to the west, Ondo, Edo, Anambra, and Enugu to the south, and Benue to the east. Its approximately 3.3 million people (2006 census) are predominately Igala, Ebira, and Okun, although there are also people of Bassa, Nupe, Ogugu, Gwari, Kakanda, Oworo, Ogori and Eggan descent present in the state. Agriculture is the mainstay of the state's economy, with the principal crops including coffee, cocoa, palm oil, cashews, groundnuts, maize, cassava, yam, and rice (www.kogistate.gov.ng). Mineral resources include coal, limestone, iron, petroleum, and tin.

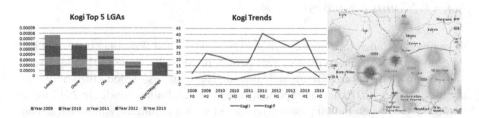

Fig. 8.5 Bar Chart shows annual per capita incidents for top LGAs; Line Graph shows trend in # of incidents and fatalities in the state; Heat Map shows hotspots of violence from 2009 to 2013—Nigeria Watch data formatted and uploaded to Peace Map (www.p4p-nigerdelta.org/peace-building-map)

The fact that the state is bordered by ten other states and is the main gateway to the north makes it a strategic corridor for issues of peace and security in Nigeria. The most common type of violence in the state is gang violence and criminality such as muggings, kidnappings, carjacking and armed robberies. However, between 2009 and 2013, Kogi state was one of the least violent states in Nigeria.

Terrorism and religious violence occurred in Kogi in 2012, when gunmen killed at least 19 worshippers at a Christian worship center near Okene area, and suspected insurgents stormed a prison freeing 119 inmates. There were also some clashes between suspected insurgents and security forces.

The first governor of Kogi in the Fourth Republic was Abubakar Audu of the ANPP, who was elected in 1999. In 2003, Ibrahim Idris was elected on the PDP platform, but his election was annulled in 2008. Speaker of the House Clarence Olafemi (PDP) took over as acting governor until after the redo of the election that Idris won a second time. In 2012 Idris Wada, also of the PDP, took the governorship after winning the election. The following LGA-level breakdown describes risk factors in the five most violence-prone LGAs in the state. The incidents described are not exhaustive but give a brief overview of the scope and tenor of the types of issues reported in the data during the five year period of 2009-2013.

8.4.1   Lokoja

Lokoja LGA lies at the intersection of the Benue and Niger rivers. It had the most reports of violence per capita of all the other LGAs in the state, but considering the low levels of violence in the state overall, the severity was not particularly high, though the types of issue reported cut across a wide variety of categories. In 2009 there were reports of criminal violence and a student protest over the lack of compensation to teachers. In 2011 there was some political tension reported prior to the general election. In 2012 there were protests over the partial removal of the fuel subsidy. A prison break was reported in which one of the security officers was killed. A bomb was detonated outside a church. Perpetrators were suspected to be Islamist insurgents. In Nigeria's most severe rainy season in decades, dozens were reportedly drowned in accidents associated with the flooding. In 2013 several were reportedly killed in local government election violence. Separately, teachers protested the lack of payment of the previous month's salaries. Several armed robberies and other murders were also reported during the year.

8.4.2   Okene

In 2009 tensions were reported within the PDP over financial issues, which led to riots killing two. Separately, masquerade activities were banned in Okene due to the violence that often accompanied the tradition. In one case, however, a masquerade did take place which led to violence. In 2010, two people were reportedly killed in an inter-communal clash. In 2011 there was post-election violence reported. Several people were also killed in a series of armed robberies. In 2012, clashes between security forces and insurgents associated with Boko Haram killed about ten. A few months later, gunmen stormed a church, killing 19. Three people were also killed inside a mosque. In 2013, several were killed by an improvised explosive device, planted by suspected Islamist insurgents.

8.4.3   Ofu

In Ofu LGA, the majority of incidents reported throughout the year had to do with political violence. In 2009, two people were reportedly killed in political violence between PDP and ANPP after the court upheld the gubernatorial election results, the process of which had been fraught with confusion. Political tension continued into 2010, when clashes between political thugs reportedly killed about ten in the months of February, May, and August. In 2011, an election year, political clashes were reported in April and December. In 2012 a shootout was reported between youths, also related to political differences. In 2013 a couple murders were reported. The former governor's brother was reportedly kidnapped, in what was suspected to be a politically motivated crime.

8.4.4   Ankpa

In 2009 one person was killed in a shootout between police and a gang. In 2010 teachers protested over compensation issues. Several robbers ambushing travelers at a roadblock were reportedly killed by police. In 2011 a college was closed after violent demonstration and the reported death of several people suspected to have been killed by cultists. In 2012, several people were reportedly killed in a bank robbery. In 2013 several people were reportedly killed in election violence, including a PDP candidate. Separately, two armed robbers were reportedly killed by police.

8.4.5   Ogori/Magongo

Ogori/Magongo has by far the smallest population of any of Kogi's LGAs. The only significant incident reported in this LGA was in 2012, when gunmen attacked a police station, killed a corporal, and stole some weapons.

Chapter 9
Conclusion

In years past, the collection, coding, integration, and analysis of conflict data as done here, would have been exceedingly cost prohibitive to non-profits such as The Fund for Peace. However, due to advances in technology and effective collaboration among local, national, and international stakeholders, situational awareness at multiple levels of granularity is now possible, even in countries like Nigeria, emblematic of complexity. This relatively new ability to analyze trends at multiple levels of analysis simultaneously is critical for a better understanding of the conflict landscape. Sometimes it is necessary to peel back layers so as not to misread the big picture and sometimes the big picture is necessary to understand why a particular incident may have taken place. It is tempting, sometimes to think that the closer you get to the ground, the closer you get to the truth. But this premise is belied when a colleague is killed and there is no clear answer as to what extent the killing may have been triggered by ethnic, communal, political, criminal, and/or interpersonal factors. At the other extreme, an aggregation of data at the national level tells you very little about the intermediate conflict ecosystems and how they do or do not interrelate in a given time period. A bombing in Kaduna by Boko Haram might mean something very different by way of perpetrator, objective, and effective response, than a similar bombing in Maiduguri. Also, localized pastoral conflicts in the Middle Belt or communal violence in the Niger Delta may or may not be influenced by broader political and sectarian dynamics. As stakeholders, regardless of mandate, whether local, sub-national, or national, effective peace and security planning requires this multi-level analysis.

However, a reality check may be in order here. Maps, graphs, and tables, such as those in this book, shed light on patterns and trends of violence and conflict risk. But they do very little with regards to clarifying the causes of and solutions to those conflict patterns. For that, a new level of analysis is required. Stakeholders must examine the patterns and trends, and then undertake a deep qualitative assessment of the social, economic, political, and security drivers to understand the causes. Then they must perform a careful scoping and Strengths, Weaknesses, Opportunities, and Threats (SWOT) analysis as part of a planning exercise for the solutions.

© Springer International Publishing Switzerland 2015
P. Taft, N. Haken, *Violence in Nigeria*, Terrorism, Security,
and Computation, DOI 10.1007/978-3-319-14935-6_9

Indeed, this type of participatory qualitative engagement, using the quantitative data aggregated in this book, is happening at various levels in Nigeria. Since 2010, the Institute of Human Rights and Humanitarian Law, led by Anyakwee Nsirimovu, has been convening meetings with local civil society groups every several months in cities such as Port Harcourt, Jos, and Kaduna, to do exactly that. More recently the Foundation for Partnership Initiatives in the Niger Delta (PIND) has been proactively facilitating this process through their Partners for Peace initiative. First they supported the development of the mapping tool (http://www.p4p-nigerdelta.org/peace-building-map) for the integration of data from as many existing conflict assessment and early warning initiatives that were willing to collaborate. Then they launched a grassroots network in all nine Niger Delta States so that traditional rulers, women's leaders, youth leaders, civil society, private sector actors, and government officials could analyze that data and plan their own, locally owned and locally driven response. PIND, as a foundation, is there to provide guidance and in-kind support to fill gaps and ensure that those plans are ultimately successful. For example, in 2014 the P4P Cross River State Chapter did an intensive analysis of the data included in this book and agreed that Abi LGA (see p. 40) was a priority area for conflict mitigation. Upon doing a qualitative assessment, using FFP's Conflict Assessment System Tool (CAST), they then planned an intervention in Ediba and Usumutong to mitigate intercommunal land conflict issues. Meanwhile, at the national level, donors and practitioners have formed a Peace and Security Working Group, to share information, including data and analysis integrated onto the Peace Map, and to coordinate efforts for conflict mitigation. The Fund for Peace has been involved in all of the initiatives described above.

To say that conflict is complex is not particularly new or insightful. However, grappling with that complexity for the promotion of peace and security is easier said than done. It requires the collaboration of local, national, and international stakeholders, including development, security, and governance actors, for participatory conflict assessment and effective response. Collaboration can be fraught with difficulty, especially in polarized societies. The priorities and imperatives of the security, development, and governance sectors are not always or immediately in alignment. There can be competing interests, or a lack of trust. Data can be misinterpreted or misused. But if you accept that multi-stakeholder collaboration is the only hope for success in the promotion of sustainable peace, then the development of publically available ICT infrastructure such the Peace Map in support of social infrastructure (P4P, WANEP, PSWG, UNLocK, etc.) is the way forward.

Nigeria has many challenges. But it has a lot going for it. The human and social capital in Nigeria is robust. Civil society is deeply engaged. The private sector is energized. There is reason for hope.

Bibliography

Action on Armed Violence. (2014, December 12). The Violent Road: Nigeria's North East. Retrieved from http://aoav.org.uk/2013/the-violent-road-nigeria-north-east/

africajournalismtheworld. (n.d.). Africa News and Analysis. Retrieved from http://africajournalismtheworld.com/

Asuni, J. (2009, September). Understanding the Armed Groups of the Nigeria Delta. Council on Foreign Relations Working Paper.

Baker, P. (2012, September). Getting Along: Managing Diversity for Atrocity Prevention in Socially Divided Societies. Retrieved from http://www.stanleyfoundation.org/publications/pab/BakerPAB912.pdf

BBC news Africa. (2012, January 5). Nigeria fuel subsidy: Police tear gas Kano protesters. Retrieved from http://www.bbc.com/news/world-africa-16425111

Campbell, J. (2013, July 18). Nigeria's Civilian Joint Task Force. Retrieved from http://blogs.cfr.org/campbell/2013/07/18/nigerias-civilian-joint-task-force/

CIA World Factbook. (2014). Retrieved from https://www.cia.gov/library/publications/the-world-factbook/geos/ni.html

Elbagir, N., & John, H. (2012, January 21). Scores dead as assailants target northern Nigerian city. Retrieved from http://www.cnn.com/2012/01/21/world/africa/nigeria-explosions/index.html?hpt=hp_t3

European Commission, Humanitarian Aid and Civil Protection. (2014, September). Nigeria Echo Fact Sheet. Retrieved from http://ec.europa.eu/echo/files/aid/countries/factsheets/nigeria_en.pdf

Falola, T., & Matthew, M. H. (2008). A History of Nigeria. Cambridge University Press.

Francis, P. (n.d.). Retrieved from http://www.irinnews.org/report/88906/analysis-nigeria-s-delta-amnesty-at-risk-of-unravelling

Francis, P. a. (2011). Securing Peace in the Niger Delta. Retrieved from http://www.wilsoncenter.org/publication/securing-development-and-peace-the-niger-delta-social-and-conflict-analysis-for-change

Haken, N., Taft, P., & Jaeger, R. (2013). A CAST case-study: assessing risk in the Niger Delta. In Handbook of computational approaches to counterterrorism, ed. V.S. Subrahmanian, Springer.

Haken, N., Carreira, F., Egorova, E., & Hersh, R. (2012, December 10). Nigeria: Beyond Terror and Militants. Retrieved from http://library.fundforpeace.org/cungr1215

Hanson, S. (2007). MEND. The Niger Delta's Umbrella Militant Group. Retrieved from http://www.cfr.org/nigeria/mend-niger-deltas-umbrella-militant-group/p12920

© Springer International Publishing Switzerland 2015
P. Taft, N. Haken, *Violence in Nigeria*, Terrorism, Security, and Computation, DOI 10.1007/978-3-319-14935-6

Hilton, K., & Ndukong. (2013, July 31). Nigeria: Scores Die In Kano Bomb Blasts. Retrieved from http://allafrica.com/stories/201307311236.html

Human Rights Watch. (2005, May). Revenge in the name of religion: Thy cycle of Violence in Plateau and Kano States.

Murphy, D. (2014, May 6). 'Boko Haram' doesn't really mean 'Western education is a sin'. Retrieved from http://www.csmonitor.com/World/Security-Watch/Backchannels/2014/0506/Boko-Haram-doesn-t-really-mean-Western-education-is-a-sin

Musa, I. (2012, June 11). White paper on 2011 Post-Election Violence: Divide Kaduna state for peace to reign.

Nigeria: Post-Election Violence Killed 800. (2011, May 17). Retrieved from http://www.hrw.org/news/2011/05/16/nigeria-post-election-violence-killed-800

Sergie, M. A., & Johnson, T. (2014, May 3). Boko Haram. Retrieved from http://www.cfr.org/nigeria/boko-haram/p25739

Ubhenin, O. (2013). Edoror The Federal Government's Amnesty Programme in the Niger-Delta: An Appraisal. Retrieved from http://acikerisim.lib.comu.edu.tr:8080/xmlui/bitstream/handle/COMU/550/Oscar_Edoror_Ubhenin_Makale.pdf?sequence=1&isAllowed=y

United States Department of State Publication. (2010, August). http://www.state.gov/documents/organization. Retrieved from http://www.state.gov/: http://www.state.gov/documents/organization/141114.pdf

Xinhua. (2012, January 22). 150 Killed in Attacks in Northern Nigeria State. Retrieved from http://english.cri.cn/6966/2012/01/22/2724s677504.htm

Printed in the United States
By Bookmasters